TRACKING THE HIGHLAND TIGER

TRACKING THE
HIGHLAND TIGER

Marianne Taylor

BLOOMSBURY WILDLIFE
LONDON · OXFORD · NEW YORK · NEW DELHI · SYDNEY

For TT

BLOOMSBURY WILDLIFE
Bloomsbury Publishing Plc
50 Bedford Square, London, WC1B 3DP, UK

BLOOMSBURY, BLOOMSBURY WILDLIFE and the Diana logo
are trademarks of Bloomsbury Publishing Plc

First published in Great Britain 2019

A catalogue record for this book is available from the British Library.

Library of Congress Cataloguing-in-Publication data has been applied for.

ISBN: HB: 978-1-4729-0092-0; ePub: 978-1-4088-4543-9;
ePDF: 978-1-4729-7025-1

2 4 6 8 10 9 7 5 3 1

Typeset in Bembo Std by Deanta Global Publishing Services, Chennai, India
Printed and bound in Great Britain by CPI Group (UK) Ltd, Croydon CR0 4YY

To find out more about our authors and books visit www.bloomsbury.com
and sign up for our newsletters

Contents

Prologue

The veterinary nurse took the empty cat carrier from me and, I think, darted me a quick look of sympathy as she turned away. I chose a chair in the hallway and watched her disappearing into a side room.

I began to read the notices on the walls, about neutering and fly-strike and lungworm, but didn't get very far as the nurse was back a moment later. She handed me the carrier – it felt barely heavier than before and there was no shifting of weight as I balanced it in my arms, nothing to give any sense of a living thing inside. 'Good luck,' the nurse said. 'She's OK if you can get hold of her – she doesn't bite. She's just very scared.'

As I headed down the short hill to my home I spoke into the cat carrier, through the narrow vents in its plastic walls. I described the parked cars and garden flowers as we passed them. I said it would all be OK, in my softest tones. The almost-weightless animal in the carrier didn't move. I imagined her frozen in fear, cowering under the blanket I'd put in there to soothe her.

Back home, I set the carrier on the floor, took off its top, and sat down on the sofa. All that I could see was a rumpled pile of blanket. I carried on talking and, a few minutes later, the striped folds of fabric shifted and the kitten poked her head out, turned to meet my gaze. Her face was a flower, her stare wide, intense, horrified. My words dried up; I was transfixed and appalled. A lifetime of having pet cats hadn't prepared me for this. The air seemed charged with the force of her distress – a wild thing, panicking in a trap. Then she hopped out of the carrier and made for a cracked-open cupboard. Her little banded tail whisked out of sight around the cupboard door.

I didn't see the kitten again for about three days. At night she left her hiding place to eat and use her litter tray. By day she waited, still and silent, out of view, and I would sit and talk quietly and endlessly to what may as well have been an empty room. The notion of getting hold of her, as the vet nurse had suggested, was laughable. I doubted I would ever even touch her.

Slowly – very, very slowly – I began to see her. At first unexpectedly, a little dusky shape darting behind a bookcase or under a bed when I entered a room. Then, after half an hour of flicking a knotted bit of string across the floor close to where she was hidden, a paw reaching out to hesitantly pat at the ground. It was weeks, though, before she had become brave enough to let me see her clearly. She was a stocky little animal, round-faced. Her fur was grey-brown, marked boldly with black swirls. It occurred to me that, had her pattern been of mackerel-stripes, she would have been the image of a true wildcat. As it was, I took to calling her my little wildcat – my *gatita fiera* – anyway, though her real name was Sookie. She would never be like any other pet cat I'd known.

The domestic cat descends from the African wildcat, *Felis lybica*. Its very close cousin is the European wildcat, *Felis silvestris* – 'cat of the woods'. Just as we domesticated wild wolves and from them used selective breeding to produce the fabulous array of domestic dog breeds that exists today, so we gradually turned wildcats into tame cats.

But it doesn't take much to turn the process backwards, and a tame cat's kitten can very easily become a wild cat – or at least a feral cat, fearful of humans and only suited to a life of wildness and self-sufficiency. Born in the wild to a stray, and untouched by any human hand until she was trapped by a cat-rescue charity at five weeks old, my kitten had already missed the start of her crucial 'socialisation window', within which she would, if handled regularly, have become unafraid of people. Maybe it wasn't quite too late, but for the next seven weeks she cowered in a cattery pen because no one had time to try to tame her. Then she came to me, three months old, at the end of her socialisation window and no closer to tolerating human contact. A project to tackle. Could I de-wild this feral kitten at all?

The weeks passed with slow progress. The biggest leap forward came when she had been with me a month, and the rescue charity brought me a second kitten, a confident and fully socialised male a couple of weeks her junior. He ignored her panicked hisses and befriended her through sheer force of will. She watched his confidence with me, and her own grew, slowly and in fits and starts, but it was not too long before she began to permit very slow, careful strokes on the head.

The two kittens were adopted permanently by friends of mine – they saw Sookie's sweetness under the shyness. She lived with them happily for seven years. As her fear receded further, her good-natured character came to the fore. When my friends went away for a year of adventurous travelling, they asked me to take care of her again.

It's now more than eight years since I brought that petrified kitten home. As I write, she's with me for one more week and then she returns home. The change in her has amazed me. She purrs and hurries to me when she sees me. I can stroke her, lift her up, she'll curl up on my lap, occasionally she bumps her nose against mine. In many ways, she is like any other friendly pet cat. But the original wildness in her has not gone away – it is no longer obvious all the time but it remains at the heart of what she is. If I move too quickly or carelessly that old glaze of instinctive fear overtakes her eyes and she's gone, making a low-backed sprint away to any place I can't reach her, without a backward glance. A knock on the door, an unfamiliar voice, car tyres crunching on the gravel outside – all of these send her running too.

Some would say that even the cuddliest pet cat retains some grain of this wildness. Those original wildcats that, one way or another, became domestic cats, were African wildcats – *Felis lybica*. The African wildcat is, in superficial ways, a rather different animal to the legendary

Scottish wildcat *Felis silvestris* (*grampia*), the most elusive and threatened wild mammal to haunt the British countryside. But appearances deceive, and the two are separated more by miles then genes. Get them together, and they are fully, disastrously compatible. They breed together and the Scottish wildcat's genetic uniqueness is watered away. Not only that, but they compete for the same resources and share the same sicknesses.

Most of us who have pet cats truly love them and care for them accordingly, but that grain of wildness they hold – that self-sufficiency and survivability – makes it easy for us to be careless at times. Be negligent with tame cats and they or their young may turn into feral cats, living wild and off their wits. Because they carry this potential, domestic and feral cats today threaten to eradicate true wildcats everywhere – but especially in Scotland.

Of all the myriad ways there are for a wild animal to disappear from the world, death at the paws – or genes – of a domestic cousin is one of the most unusual. In truth, Scottish wildcats had already been laid low by other causes and conflicts before the modern proliferation of feral and free-roaming domestic cats. The same old threats and enemies still exist today, but not to anything like the extent that they once did – it is interaction with not-wild cats that threatens to deliver the *coup de grâce* to the tiny remnant population of our own native wildcats – a tragedy unfolding before our eyes and a fiendishly intractable problem to try to solve. Sookie the once-feral kitten, now a nearly-tame cat, is a fragment of this story. She was saved, but another million or so feral cats in Britain today live out their difficult, brief and violent lives as essentially wild animals. They are the innocent victims of our carelessness, as much as the Scottish wildcats are, but it is the wildcats that stand on the brink of obliteration.

Many say it's already too late. Some say that it doesn't even matter, that a cat is a cat is a cat – and if feral cats replace wildcats, have we really lost anything? This attitude is a devastating threat in itself. Others, though, will not accept that the unique and marvellous Scottish wildcat is a lost or pointless cause. Through their efforts and ingenuity in the lab and the zoo and the wild, sparks of hope give light and life to a real and lasting future.

This book traces the steps of the wildcat through a bloody and troubled history, through the struggles of the present and into the unfolding future.

CHAPTER ONE

A Cat in Context

It's one of the first words most of us British-born kids master when we are tiny. Even if there isn't one in our own home, we'll still know a cat when we see one because they are ubiquitous. Like nearly all domestic animals, cats are highly variable, but their essential felinity is evident over and above differences in colour, pattern, fur length and build.

When we're a little older, we start to learn about other cats that share our world – the wild ones. The big ones. We learn about lions and tigers, leopards and jaguars, servals and ocelots. We learn that they are cousins to our pets, and we see the cat-ness in them, in their beautiful patterned faces and glowing eyes, their whiskers and restless tails, lithe grace and ability to hunt and to kill. That sets us to wondering how long we might survive if our house cat grew to the size of a lion. We begin to recognise and respect the wildness in the tame cat.

There are about 40 different species of cats in the world, most of which are rare, obscure or both, and the chances of any of us clapping eyes on more than half a dozen cat species in the wild are pretty low. I have one extremely well-travelled, wildlife-obsessed and well-off friend who's seen more than 20. My own tally is a rather more dismal at two: lions in southern Africa, and an opportune encounter with Iberian lynxes in Spain. (Or possibly I should say it is two and a bit, but more about that later.)

Here are the wildcat species living in the world today, according to a 2017 revision by the IUCN (International Union for Conservation of Nature) SSC (Species Survival Commission) Cat Specialist Group. However, please note that taxonomists – those whose job it is to work out the relationships between different kinds of living things – are continually updating their classifications in light of new research (especially genetic evidence), and this list may be out of date by next week.

Lion *Panthera leo*
Jaguar *Panthera onca*
Leopard *Panthera pardus*
Tiger *Panthera tigris*
Snow leopard *Panthera uncia*
Sunda clouded leopard *Neofelis diardi*
Clouded leopard *Neofelis nebulosa*

African golden cat *Caracal aurata*
Caracal *Caracal caracal*
Serval *Leptailurus serval*
Pampas cat *Leopardus colocola*
Geoffroy's cat *Leopardus geoffroyi*
Güiña *Leopardus guigna*
Southern tiger cat *Leopardus guttulus*
Andean mountain cat *Leopardus jacobita*
Ocelot *Leopardus pardalis*
Northern tiger cat *Leopardus tigrinus*
Margay *Leopardus wiedii*
Borneo bay cat *Catopuma badia*
Asiatic golden cat *Catopuma temminckii*
Marbled cat *Pardofelis marmorata*
Canada lynx *Lynx canadensis*
Eurasian lynx *Lynx lynx*
Iberian lynx *Lynx pardinus*
Bobcat *Lynx rufus*
Puma *Puma concolor*
Cheetah *Acinonyx jubatus*
Jaguarundi *Herpailurus yagouaroundi*
Leopard cat *Prionailurus bengalensis*
Sunda leopard cat *Prionailurus javanensis*
Flat-headed cat *Prionailurus planiceps*
Rusty-spotted cat *Prionailurus rubiginosus*
Fishing cat *Prionailurus viverrinus*
Pallas's cat *Otocolobus manul*
Chinese mountain cat *Felis bieti*
Jungle cat *Felis chaus*
African wildcat *Felis lybica*, and including the domestic
  cat *F. l. catus*
Sand cat *Felis margarita*

Black-footed cat *Felis nigripes*
European wildcat *Felis silvestris*, including the Scottish
wildcat *F. s. grampia*

You might also know some of these cats by their
alternative English names – for example, the güiña is also
known as the kodkod, and the northern tiger cat also
goes by the names oncilla and little spotted cat. The
puma is another with many names – its best known in
the English language are cougar and mountain lion, but
its other monikers include painter, panther and catamount.
All these names reflect its extensive distribution, crossing
dozens of linguistically diverse countries, as historically
its range encompassed all of South America and almost
all of North America too.

These animals make up the cat family, Felidae; one of
about 16 families that make up the order of mammals
called Carnivora – the meat-eating mammals. Carnivores,
as a group, make their living by hunting, killing and
consuming other animals. Many are omnivorous, and a
few individual species have adapted to a plant-dominated
diet – most famously the giant panda, a bear that subsists
almost entirely on bamboo. However, as a group, they
are killers and meat-eaters, and this fact is evident in the
structure of their teeth, jaws, feet, sensory systems and
digestive tracts.

Carnivora splits neatly into two subgroups: the
feliforms ('cat-shaped') and Caniforms ('dog-shaped').
Many caniforms are omnivores and are adapted to run
on the ground rather than climb – their claws tend to be
non-retractile. Their teeth are less specialised but more
numerous to deal with a more varied diet. Caniform

carnivores outnumber the feliforms in terms of species diversity, though the number of families is about the same. Caniform carnivores comprise the weasels, badgers, otters and their allies (the mustelids – Mustelidae), and also the bears (Ursidae), the raccoons (Procyonidae), the skunks (Mephitidae) and the seals (Phocidae and three other families). The subgroup is completed by the true dogs and the foxes (Canidae).

The feliforms are more strictly carnivorous, and they have shorter and stouter jaws with large fangs and carnassial teeth, for a stronger bite and better ability to tear meat. Many of them dwell in trees and have dappled patterns to disguise them as they hide among foliage. Most can pull back or sheath their claws (which means the claw-tips retain sharpness for climbing and grabbing). Besides the cats themselves, this grouping includes the family Viverridae, made up of the genets and the African linsang (little, spotted tree-dwellers of almost-supernatural grace), the more terrestrial civets and the curious binturong or 'bearcat', a big long-tailed beast with a mellow disposition and (unusually for a feliform) a truly omnivorous diet. Other feliform families are the gregarious, inventive mongooses (Herpestidae) and the superficially more dog-like hyenas (Hyaenidae). The other, less well-known feliform families are Nandiniidae (with just one species, the African palm civet), Prionodontidae (the Asiatic linsangs – two species) and the unusual carnivores of Madagascar (Eupleridae – the fossa and seven other mongoose-like species).

Felidae is a large and successful family. You'll find examples of them in North and South America and throughout Europe, Asia and Africa. The family can be divided into two subfamilies, with most species forming the subfamily Felinae – the 'small cats' (although some of them are not so small). The other, much less diverse feline subfamily is Pantherinae, the 'big cats'. There are just seven species: the lion, tiger, jaguar, leopard and snow leopard (all members of the genus *Panthera*) and the two species of clouded leopards (genus *Neofelis*). It may come as a surprise to you that two species are missing from the 'big cats' subfamily: the cheetah and the puma. They are big, after all. Pumas outweigh leopards, on average, and both pumas and cheetahs easily outweigh both of the two clouded leopard species. But as regards their evolutionary pathway, pumas and cheetahs both belong within the 'small cats' subfamily and – despite their very different looks and ways of life – they are in fact sister species to each other.

An old name for the big cats was 'roaring cats', with the smaller cats known as 'purring cats'. However, only four of the five *Panthera* cats can roar: the lion, tiger, leopard and jaguar. They are able to produce their resonant rumbles thanks to a quirk in the hyoid, a horseshoe-shaped bone that sits in the throat and is tied by muscular links to the larynx. In the roaring cats, the hyoid is not fully ossified but has some tendon stretch to it, meaning that the back of the mouth can open up more widely and produce that spine-chilling sound. The snow leopard lacks this trait and was therefore thought to be not closely related to the roarers. You'll find it listed under a different name (*Uncia uncia* instead of *Panthera uncia*) in older books, yet research into big-cat genetics has revealed that it is a true *Panthera* after all (albeit the most basal or 'primitive' member of the group). Another pantherine trait lies in the eyes: they have round pupils while those of felinid cats are typically slit-shaped (there are exceptions to this, though).

The subfamily Felinae is where all the other cats on Earth belong. There are 12 genera within this subfamily, and there has also been research into their genetics since the turn of the century. This has revealed that the group is made up of seven distinct lineages. These groupings are: the lynxes (including the bobcat); the ocelot and its fellow spotty, big-nosed South American cats of the genus *Leopardus*; a trio of middle-sized African species (caracal, serval and African golden cat); the bay cat of Asia and its congeners; the leopard cat of Asia and its congeners; a curious assemblage spanning Old and New World comprising the cheetah, the puma and the bizarre jaguarundi with its attenuated, snaky body; and the Old World genus *Felis*. Cats called *Felis* are the cats we know and love best as they include the domestic cat and its wild ancestor, the African wildcat, and also the European wildcat, of which the Scottish wildcat is a subspecies.

The African wildcat varies in appearance across its extensive world range – it occurs widely through Africa and in the Middle East as well. In general, it has a short, sleek coat that is light grey-brown or tawny and has a faint tabby pattern that is most prominent on the legs, face and banded tail. It closely resembles a lightly patterned domestic tabby cat, and indeed feral cats in hot climates tend to revert to an African wildcat-like appearance over the generations, with other colour forms becoming less frequent. The European wildcat of mainland Europe and south-west Asia is very similar, but a little more thickset, with a darker and bolder pattern of tabby stripes.

Those European wildcats living in more open, arid and warmer environments tend to be more sandy-toned and shorter-coated, and have a skinny, leggy look, while those living in cooler and more forested habitats are fluffier, stockier and darker, and their tabby stripes more prominent. The Scottish wildcat takes this up another level again, having a long, thick and dark coat with a very bold pattern, and a large, muscular, square-cut and stocky frame.

The other *Felis* cats are close cousins of our pet cats, of their ancestor the African wildcat, and of European and Scottish wildcats. There are four species (sometimes more, depending who you listen to) and each has its own charm. The Chinese mountain cat (*Felis bieti*), with its thick fluffy coat, has the look of a mountain-dweller. It recalls a sandy-coloured, faintly striped Scottish wildcat, and lives in China around the Tibetan plateau. The jungle cat or swamp cat (*Felis chaus*) is similar looking but much less fluffy, as befits its habitat of warm, jungly low-lying wetlands in south and south-east Asia. The adorable sand cat (*Felis margarita*), found in deserts across north Africa, west Asia and the Middle East, is the cat world's answer to the fennec fox, with oversized ears, the most charming dainty and outraged-looking face, and a nearly unpatterned, creamy-golden coat. Finally, the elusive and diminutive black-footed cat (*Felis nigripes*) of southern African savannas is a breathtakingly beautiful creature with bold black marbled markings on its tawny coat.

Regardless of their size, cats form a very distinct and coherent group in both anatomy and behaviour. It's easy for us to see the tiger in our pet tabby and vice versa.

The internet is full of amusing videos of captive big cats behaving like pet cats – head-bumping their keepers, chasing the light spot from a laser pen across the floor, trying to squeeze into cardboard boxes too small for them and so on – and of domestic cats attacking much larger potential foes with all the fearless commitment of the king of beasts. The list of physical traits shared by the various cat species is a long one. To pick out just a few:

- Cats have a powerful build for their size and exceptional flexibility.
- They walk on their toes (a 'digitigrade' gait), rather than on flat feet like a mustelid.
- They have five toes in contact with the ground on their front feet and four on their hind feet.
- The claws are naturally held in a retracted position and can be protracted (pushed out) by muscular action. The retracted claws are shielded by protective sheaths (except in the cheetah).
- They have proportionately short, somewhat flat faces, with wide and squarish (as opposed to tapering and pointed) snouts.
- They have large eye sockets for big eyes that are adapted to provide highly movement-sensitive vision that works best in low light.
- The irises are quite light-coloured, often golden or green.
- The ears are large and mobile.
- The tongue has hard projections (papillae) on its upper surface, giving the 'sandpaper lick' we know from our pet cats. This rough tongue can rasp meat from bones and also helps with thorough grooming.

- There are well-developed whiskers around the mouth and above the eyes.
- They have 30 permanent (adult) teeth – the upper jaw has six incisors, two canines, six premolars and two molars, and the lower jaw is the same but with only four premolars, i.e. I3/I3, C1/C1, P3/P3, M1/M1 and I3/I3, C1/C1, P2/P2, M1/M1.

Most wildcat species also have well-marked coats – the most common is a tawny or yellowish backdrop with an overlaying pattern of black or dark-brown spots, but a fair few are striped rather than spotty, including our own Scottish wildcat. However, some of those that hunt in more open or more watery environments have almost unmarked coats – among them are the lion and the puma, the caracal and the jaguarundi. Likewise, those populations of African and European wildcats that live in drier, sparsely vegetated environments are much less well marked than those that inhabit forests. Coat colour and pattern offer camouflage, and so are absolutely vital for a hunter relying on stealth. An unintended consequence of all those beautifully patterned coats is that cats of all kinds are hunted for their fur, both legally and illegally, which presents a serious threat to the survival of some species.

Cats are also prone to melanism – excess melanin pigmentation – giving a black or darker-than-usual coat. This arises from a genetic quirk, and is quite clearly a problem for some cat species. A black lion in eastern Africa, for example, would probably not fare too well trying to hunt by stealth in its tawny-golden savanna surroundings. However, in some environments a black coat can actually offer good camouflage, meaning that melanistic animals survive well and the genes they carry can persist in the population. For example, melanistic leopards ('black panthers') are relatively common in the dense forests of mainland Malaysia. The jaguarundi, an unspotted and unstriped cat of wet swampy grassland and forest, occurs in a blackish-grey colour morph as well as in a puma-like reddish-brown one. Darker forms are more numerous in tropical rainforest, whereas paler ones predominate in more open habitats, but the cats themselves make no distinction and the two forms readily interbreed.

Behavioural traits are a bit harder to pin down than physical ones, but it is fair to say that cats, whether big or small, are well adapted to a particular style of hunting that Spike Milligan once summed up beautifully in his short poem 'Tiger, Tiger Burning etc'.

The slinking feline stalk is practised by all cats, from tigers to the tiny rusty-spotted cat that is barely half the size of a typical domestic pet cat. Cats are fast runners over short distances, with impressive acceleration, but they are not stamina athletes – they go fast and tire quickly. Because their attacking dash will not be a long one, cats need the ability to get close to their prey before they begin their race, in the hope of taking it by surprise and getting their pounce in before the victim can reach its own full speed (or fly away). This is where their stealth comes in. The creep, the wait, the fixed stare, the sudden surge – it's the same whether the quarry is a feather toy twitching on a string or an oblivious antelope. Even lions, which are true cooperative hunters, employ stealth to get close to their prey, instead of using the open, relentless pursuit tactic of those other celebrated team players, the wolves.

Another option for a skilled stealth hunter is ambush hunting. The leopard excels at this, lying in wait in low tree branches for some hapless hoofed grazer to wander below. Gravity is all that is then required, as long as the drop is accurately aimed.

Cats are efficient killers. Their instinct is to try for a firm jaw grip at the head or throat, which will suffocate the victim and quite possibly sever the spinal cord as well. A pride of lions usually dispatches their kill far more quickly than a wolf pack does – the wolves tend to chase and bite at the unfortunate victim until it keels over from exhaustion rather than from any single devastating injury. However, as many pet-cat owners know to their chagrin, cats are not above crippling their prey and then playing with it. If the need to eat isn't urgent, this horribly cruel-looking behaviour makes sense: it's a valuable opportunity to practise hunting behaviour. Even the most intelligent pet cat cannot 'know' that it will never actually need to hunt for food in its life; the instinct to hone and build skills is compelling and natural. In chapter 3 we will take a closer look at how cats – wild and tame – have been shaped by evolution into the expert hunters they are today, from the way their bodies are put together to the way they behave.

By and large, cats live and hunt on their own. The obvious exception is the lion, which lives in social groups, with closely bonded groups of females hunting together and rearing young together. The pride also has an adult male (sometimes two or more related males), though these males are really only used as sperm donors and to help defend the cubs – most often from other males that would kill all cubs in the pride if they managed to depose the incumbent king or kings. Cheetahs also show social behaviour sometimes, although in their case the bonds are between related males. Any male cheetah who survives to adulthood alongside a brother or two is in a much better position than an 'only cub' – he and his brothers can hunt together, pursue females together and generally look out for one another. Sociality can work well in the savanna, where herds of large prey animals are abundant, and cooperation can bring down a meal big enough for a whole group. An extinct American subspecies of the lion (*Panthera leo atrox*) pursued a similar lifestyle on the North American prairies, hunting buffalo and other big hoofed mammals. In most other environments, though, quarry is smaller and scarcer, and if you are a cat, it is much easier to just fend for yourself.

Sociality in the smaller cats is nearly non-existent, though feral domestic cats do tend to form colonies – especially in places where there's a ready food supply – based around groups of females rearing their young cooperatively. The more typical cat lifestyle, however, is to live and hunt alone, to have brief and often extremely tense encounters with the opposite sex only when it's time to mate, and for youngsters to leave their mothers and strike out alone as soon as they are competent hunters in their own right.

The modern cats, then, are pretty successful in evolutionary terms, having achieved a pervasive worldwide distribution and diverged into an impressively diverse array of species. Their evolutionary story dates back many millions of years. As with all animal families, cats of all kinds – extant and extinct alike – descend from one common ancestor. The truth of this is written into their mitochondria (tiny structures that reside in each cell of their bodies and carry a set of DNA down the maternal reproductive line). Track back further into the past and you'll find the common ancestors to all carnivores, or at least you'll find their fossils. They were small, tree-climbing, weasel-like animals called miacids. The miacids' descendants diverged into caniforms and feliforms some 50 million years ago, and the first true cats appeared around 32 million years ago.

Some three or four million years ago, at about the time we humans were evolving towards our current modern form and getting to grips with life on Earth, another subfamily of cats was prowling the untamed wilds. These were the machairodonts. They appeared about 16 million years ago, and the last of them survived to some 11,000 years before today. They included the so-called sabre-toothed cats, the most famous of which were the 'sabre-toothed tigers' of the genus *Smilodon*. These were huge, robust animals, almost more bear-like than feline in their build. No doubt they came into conflict with humans, who would have been competition for them as well as potential food. Studies have shown that, despite their greatly enlarged upper canines, *Smilodon* cats would have had a pretty puny bite force compared to, say, a modern lion, although they did have formidable neck muscles that would have helped force the upper jaw down and drive the giant fangs into a prey animal's body when the mouth was wide open. Theories over how these cats hunted and how they killed continue to rage today. It is possible that they were social and took down mammoths and other large herbivorous mammals with teamwork, but we won't know for sure because the machairodonts died out completely – as did the members of Proailurinae, a subfamily of small, tree-climbing creatures that were the first 'true' cats. In contrast, Pantherinae and Felinae made it to the modern day (albeit leaving a few extinct genera behind on the way).

As the world's continents shifted about and ice ages came and went, so the ancestral cats moved with them, and natural selection acted on their gene pools to refine their bodies and minds into forms adapted to function successfully in an array of habitat types. Those that live today are success stories (for now), though a distressingly high proportion of wildcat species, both great and small, are now seriously threatened with extinction.

One of the earliest fossil cats assigned to the genus *Felis* is *F. lunensis*, also known as Martelli's cat. This species was living in Europe some three million years ago and may be a direct ancestor to today's European wildcats (and therefore the Scottish wildcat). If not, it would have been a cousin to the true direct ancestor. There is no complete specimen so it's difficult to imagine what this cat might have looked like, but it probably resembled a modern Scottish wildcat quite closely. It would surely have had a thick and insulating fur coat at least, as it was living in some pretty cold regions. The fossil record shows that true European wildcats began appearing about two million years ago. Great Britain was not an island back then but connected to the mainland, and land animals like European wildcats could and did spread unimpeded across the entire extensive European land mass, as far as their adaptability to habitat change would allow. Wildcat fossils dating back at least 500,000 years have been found in England, though as ice ages came and went, the distribution of the cats (and other animals) shifted around, southwards when the ice came and back north as it went. Wildcats need woodland, and woodland cannot endure prolonged ice cover.

The last major ice age changed everything more radically. It began about 26,000 years ago and left all of what would become the British Isles except the far south-west locked under ice sheets for at least 8,000 years. The treeline retreated as the ice advanced, and all the forest-dependent animals, including the European wildcats, were pushed far south into what would be mainland Europe, until the ice began to thaw. As it thawed, and vegetation gradually changed from being sparse and tundra-like to new forest growth, so animals spread north again, but their window for achieving this was limited. Doggerland, the low-lying expanse of land that connected modern East Anglia with the mainland, began to disappear as sea levels rose and the English Channel was born, and 9,000 years ago the 'island-ification' of Great Britain was complete. Ireland had become separated from Great Britain well before that – about 18,000 years ago.

Not all of the European land mammals returned to Britain in the short window when they could have. The brown hare is an example of one that didn't make it. It was abundant in mainland Europe up to the newly formed English Channel, but not in the British Isles until humans decided to bring it over themselves. Several others land mammals recolonised Britain but didn't reach Ireland in time – among them the weasel, the short-tailed vole and the common shrew. Whether European wildcats made it to Ireland or not is a point of contention, which we'll look into further in chapter 5. In any case, they are not there now, but they did get back to mainland Britain before the access route vanished under the sea. They also reached peninsulas that would become islands – there is fossil evidence of their presence on the Isles of Man, Skye and Bute.

Some 8,000 years ago the British Isles was heavily forested, with tree cover over about 85 per cent of its land area. All of this wildcat habitat meant that *Felis silvestris* would have fared well and was probably pretty common everywhere in those years, but the good times were short-lived. Neolithic humans were beginning to discover the benefits of keeping domestic livestock and the many uses for cut wood. The following 2,000 years saw massive forest clearance across lowland Britain. Wildcats, losing their habitat fast, were also about to take on a new and unfortunate role in the human consciousness: that of vermin. They were persecuted as thoroughly as any other animal equipped with fang and claw because they could and did take poultry and small livestock when the opportunity presented itself. By the start of the seventeenth century, there were virtually no wildcats left in south-eastern England, and their northwards retreat continued through the next 200 years. Although they could to some extent adapt to a more open landscape, the pressure from hunting was too much, and some time in the mid-nineteenth century the wildcat was eradicated from England and Wales. Britain's native cat now lived only in Scotland,

and soon only in the wildest Highlands of north Scotland. Yet even here it was not safe from hunting, from habitat loss, and from a new danger: hybridisation with domestic cats.

We look at the decline of the Scottish wildcat in more detail later on, but it is fair to say that our own Highland tiger has only existed since the later part of the nineteenth century, as this was when it no longer occurred anywhere else in Britain. For similar reasons, the European wildcat on the mainland today also has a much more limited range than it did 200 years ago. It remains quite widespread in Spain, Italy, Greece and the surrounding countries, some larger Mediterranean islands and parts of north-eastern Europe, but its overall distribution is patchy and fragmented, and several of its individual populations are small and vulnerable.

The Scottish wildcat is usually described as looking like a large, stocky tabby cat, though the most classic examples have a subtle but definite look of solidity that marks them out as distinct from any pet tabby. The base coat colour is brown or grey-brown, with a dappled, almost shimmering appearance thanks to the agouti hairs' alternating bands of light and dark coloration. Many other mammals have a similar agouti banding on some or all of their hairs, helping disguise the solid shape of the animal and so providing camouflage. The wildcat's camouflage is further enhanced by its vertical stripes, which create a banded pattern reminiscent of shadows through tall grass or sapling trunks. The face is marked with a combination of short horizontal stripes and shorter vertical stripes, with pale surrounds to the eyes and a pale chin; the whiskers are mostly white. It has greenish-yellow eyes and a dark-pink nose. The ears are relatively small and triangular, without the tufts at the tips that are prominent in some domestic breeds. The tail is thick and marked with several black rings that become more defined further away from the body, and it has a blunt black tip.

Today, many taxonomists class the Scottish wildcat as a distinct subspecies of the European wildcat – *Felis silvestris grampia*. It is, after all, quite distinctive with its large size and heavily marked thick coat, and it has been isolated from other wildcats for thousands of years. However, not all taxonomists agree that the differences are enough to class it as a subspecies. The IUCN SSC Cat Specialist Group – who supplied the full species list at the start of this chapter – does not consider it a subspecies, instead recognising only two subspecies of *F. silvestris*, namely *F. s. caucasica* (the Caucasus Mountains and Turkey) and *F. s. silvestris* (everywhere else, including Scotland). Some other authorities take a different view and recognise not only *F. s. grampia* and *F. s. caucasica* but several other subspecies: *F. s. cretensis* on Crete, *F. s. reyi* on Corsica, and *F. s. jordansi* on the Balearics. But at the time of writing, the IUCN's own Red List website does not even state that European and African wildcats are different species, instead grouping them together as *F. silvestris* and calling them collectively just 'Wildcat'. Historically, some authorities have gone the other way completely and define the Scottish wildcat as a full, distinct species in its own right: *Felis grampia*. The use of 'Scottish' in its common name is a little misleading, as it is only very recently that it has been restricted to Scotland; it formerly occurred more widely in Britain.

Some conservation bodies, therefore, call it European wildcat or just wildcat.

Here we see where things start to get tricky with our attempts to put the natural world into some sort of coherent order. Everything seems pretty straightforward at first glance. If you look at a garden bird table when there are some birds feeding on it, you will notice (even if you know nothing whatsoever about birds) that they come in several distinctly different 'sorts'. There's a robin – brown with a red breast. A flock of goldfinches – black, white, red, brown and gold in a bold and pretty pattern – all of them alike. The blue tits zipping to and fro are little and blue, white and yellow. The great tits are a bit bigger and have some black as well as blue, yellow and white. Four groups, with complete consistency within each group and absolutely no intergrades between them. While it is easy to tell those two tit species from one another, they are similar in many ways – their general look, how they behave, what they sound like, the way their colours are arranged. They look like they might be cousins, more like each other than they are like the other two.

It's usually easy to distinguish different species. It's also generally easy to see which species are more like each other and so perhaps belong together in a higher-level grouping. It's not hard to imagine that, with a bit more knowledge, all living things might be similarly easy to sort into a tidy hierarchy of relatedness. But reality is much messier than we'd like, and sooner or later our lovely neat filing cabinet of life will become a bit of a mess as we add more to it and find more and more cases that are not so clear-cut.

The advent of genetic study has revolutionised taxonomy. Before, we inferred relatedness from superficial anatomical and behavioural traits only. To be fair, this did serve us pretty well. But taking a more careful look at anatomy reveals much truth. A dolphin resembles a fish, but it has lungs, it has mammary glands, and it is warm-blooded – you don't need to delve too deeply to see that it's quite obviously a mammal wrapped up in a superficially fishy package. Nor do you need to be some kind of biological prodigy to look at a puma and at a cheetah and see the signs of cattish-ness in both of them, even though they look very different – one being spotty and stilt-legged and the other plain tawny-brown and built to leap rather than sprint. However, it has taken genetic study to reveal that our graceful, predatory dolphin's closest living relative (outside of its fellow cetaceans) is – surprisingly enough – the not-especially-graceful vegetarian hippo, and that the cheetah and the puma are close cousins rather than nodding acquaintances within the cat family.

More in-depth anatomical study has helped us tease out more tricky relationships, and behavioural studies (such as comparing the pitches of squeak of two near-identical bats living on neighbouring islands) has helped too, allowing us to recognise distinctiveness at a very subtle level. But comparing the actual DNA (both from the cell nucleus and the mitochondria) of two species gives us an objective, almost mathematical key to exactly how and when their evolutionary paths diverged. I am oversimplifying a little here, but molecular genetic study is the most elegant and effective of tools for determining the true family tree for all of life, and taxonomists have barely begun to make full use of it.

There is, however, a problem. For, although we have our objective DNA evidence, we don't have an objective scale to use for deciding exactly how much difference we need to see before we conclude that, for example, two populations of an animal should be 'split' into two separate species, or that two genera within a family are different enough that they should actually go up a level and be turned into two distinct families. We do now have a map of the Scottish wildcat's genetic code – its genome – yet we can't entirely agree on whether it should be treated as a subspecies of the European wildcat, or just as a superficially distinct subpopulation, or perhaps instead as a full, distinct species. There is no way to produce a genuinely objective scale because we are dealing with the diversity and messiness of reality. We can only take a guess at it. So our system remains rife with uncertainty and subjectivity, and this has real-world consequences.

The distinctiveness of the Scottish wildcat is important. It's of scientific interest, but it matters more in an emotional sense, and it's the same with any animal facing extinction. Losing a unique life form forever because of our own brutality, ignorance and negligence is a terrible, but in our minds, the degree of horror is magnified in proportion to how unique the life form is. The loss of the tiger from the world would be an unfathomable tragedy. It could happen in our lifetime – the threats are there and very real. The effort to ensure that it doesn't happen is massive, and yet we already have lost several subspecies of tiger. The Caspian tiger has been gone since about the 1970s, as has the Javan tiger. The last Bali tiger was killed in 1937. These were unique tiger subspecies, each with its own individual set of traits and its own history. We wiped them out, and they will never return. Yet the hurt, the shame and the loss are less intense because other tigers still exist.

If we conclude that the Scottish wildcat is no different in any meaningful way to the African wildcat and its domestic descendants, then our motivation to try to save it is going to take a hit. If we decide it is a full species, though, we will be spurred on by the knowledge that we're acting to preserve something extra special – the only mammal species endemic to Britain. In truth, when it comes to its genes, the Scottish wildcat is probably best treated as a subspecies of the European wildcat, but its distinctiveness goes beyond its genetic code. It is the only wildcat native to Britain. Long isolated from its mainland counterparts, it has evolved in appearance and habits to be quite unlike them in myriad ways. Its cultural connections to our own species – both ancient and modern, positive and negative – are complex and resonate across the centuries. The Scottish wildcat is ours, it is truly unique, and it stands on the edge of extinction because of all the devastation we have wreaked upon it and its world. It deserves our full and unreserved commitment to bring it back from the brink.

CHAPTER TWO

Wild Tales

I remember reading of Scottish wildcats in old nature books, and marvelling that such a fearsome creature could exist and look so similar to our own cuddly household pets. The Scottish wildcat, they maintained, is a beast of such extraordinary ferocity, strength and potential for aggression that it could kill a grown man. Every illustration showed the wildcat in a frozen, furious stance, back arched and bristling, ears pressed flatter than flat, eyes locked wide in a glare of blind rage, mouth wide open to reveal far more and far larger teeth than any of my own cats surely possessed. Even a small wildcat kitten, the books solemnly informed me, would be a ball of deadly rage when confronted by a human and would never – *could* never – be anything else.

The legend remains strong today. One rainy afternoon I was at Wildwood in Kent, a small zoo where native British animals (both extant and extirpated) are kept in roomy, natural enclosures. Having admired bears and boars, wolves and weasels, I finally tracked down the pleasingly huge and well-appointed wildcat enclosure and watched the two cats within – one dozing on a high platform and the other wandering around the cage floor. 'They're the most dangerous animals here,' remarked a man standing nearby, apropos of nothing. 'Worse than wolves, worse than lynxes.' He pointed at the prowling wildcat, which paused and gazed back at him, its eyes inscrutable. 'The keepers won't go in with them. Not safe. Untameable. That could kill a grown man, you know.'

I just smiled and nodded, but the old books came to mind and I watched the wildcat and wondered whether it really would fly at any human that came near it, determined to kill, and whether the keepers really did have to manage the cats' needs without ever entering their enclosure. The man wandered off and, a few minutes later, a petite young woman in keeper's clothes arrived and promptly answered all of my questions by opening the enclosure door and stepping inside. The sleeping wildcat didn't react, but the one pacing the floor did – dramatically. It rushed to the keeper and threw its stripy flank against her legs in a passionate show of affection. The keeper stooped and lifted up the cat, which snuggled down in her arms, butted its face against hers, and slowly cycled its front paws in the air in obvious and – to me as a lifelong pet-cat owner – very familiar delight.

I was not too sure what to make of this. I'd seen quite a few other wildcats in zoos prior to my day at Wildwood, and they'd all looked fierce enough, but this one was doing exactly what the books said wildcats never did. I thought back to *Ring of Bright Water*, that classic of nature writing by Gavin Maxwell. Although the book was primarily focused on Maxwell's pet otters, it also briefly featured a young wildcat kitten that was attempting to swim across a sea loch when Maxwell, crossing the loch by rowing boat, found it and fished it out, shutting it in a conveniently empty hamper and bringing it home. The kitten seemed accepting of its fate at first but soon became unhappy, then angry. It escaped from its prison and climbed as far up the chimney as it could, necessitating a second rescue. Encouraged back into the room from above by Maxwell's accomplice, it took up a defensive position on a high table and turned to face its would-be captor. It had transformed from a sweet-looking kitten to a 'noble, savage wild animal at bay'.

And that was only a kitten. We learned nothing more about this particular kitten – it was somehow wrangled back into its box and then passed on to a new owner, who wanted to truly test the notion that wildcats are untameable. But that passage had lived in my memory for years, and I couldn't square it with this full-grown wildcat enjoying a good old cuddle in human arms.

Just before the keeper set the cat gently back down on the ground, I took a photograph of the two of them as evidence against the next person who tried to tell me that wildcats were untameable. I talked to the keeper when she came out of the enclosure – a difficult task in itself because the over-affectionate wildcat was desperate to leave with her – and learned that this particular cat (a female called Isla) had known human love and care since her first hours on Earth. Bred in captivity, her inexperienced mother abandoned her at birth and the keeper had hand-reared her. She grew up as tame as any hand-reared domestic kitten and deeply bonded to her keeper.

I thought about my own little wild cat – my *gatita fiera* – born feral and rescued from a life in the wild at five weeks old, and not properly socialised until she was 12 weeks old. Her wildness would never completely leave her now. Feral kittens are tamed pretty easily if you can get them young. Four weeks old and it's a breeze. Even up to eight weeks old it is usually perfectly doable, but thereafter things get more difficult and if your kitten has reached 12 weeks of age the challenge is close to insurmountable. Yet my kitten was never fierce, only fearful, and the tales of ungovernable ferocity that surround the Scottish wildcat don't seem to be applied to the African wildcat or even to other European wildcats.

Watching Isla the not-wild Scottish wildcat, and talking to her keeper, made me wonder again about the kitten Gavin Maxwell had rescued. We never knew how old it was, but was there really any reason to think that Scottish wildcat kittens couldn't be tameable if caught early enough? I later read about another hand-reared Scottish wildcat kitten who also grew up tame and affectionate – more about her later – and yet legend insists this can't happen. From birth to death the Scottish wildcat's ferocity has long been its most celebrated, and feared, trait.

Today, with our Highland tiger so close to its vanishing point, another trait has taken over. Most of us who are interested in wildcats have never seen one in the wild and never will, and when we think of them we picture something not in-your-face ferocious but elusive to the point of invisibility. The change has been in our actions and our intentions, as well as in the wildcat's fortunes. We want to see a wildcat, watch it moving through its wild world from a respectful distance. It's nearly impossible because their population is vanishingly tiny now. But in times gone by wildcats were more common and would have been a threat to small livestock. Consequently many of the people who regularly encountered them would have had a different intention towards them: to hunt, trap and kill as many as possible. It's hardly surprising, therefore, that the cats they saw were all teeth, talons, spit and rage – these cats had been chased and cornered and were desperate, fighting for their lives. Many of those whose skins ended up as taxidermy specimens were posed in full attack mode, padded out to increase their size, and even fitted with giant false fangs to exaggerate the impression of savagery.

Its ostensibly fierce character and apparent refusal to be tamed made the wildcat a feared animal, but also inspired respect for the power it held in its body and spirit. Two thousand years ago, Caledonia (as the Romans called Scotland) was occupied by indigenous people known as the Cattani – a probable reference to wildcats. By the seventh century AD, Scotland was divided into seven Pictish kingdoms, and the most northerly of them,

Cait, took its name from the wildcat. Today the name has evolved into Caithness, and the region itself is still inhabited by wildcats. The fighting skills of the cat were much admired and the animal itself became a sort of totem to various warlike peoples of the region. Indeed, its image lives on in a variety of clan crests today. One such is the Clan Mackintosh crest, which bears an image of a wildcat standing tall on its hind feet, forepaws aloft and mouth open wide. It bears the motto 'Touch not the cat bot [without] a glove'. On Clan MacGillivray's crest the wildcat is seated with one clawed paw held up, and the motto is 'Touch not this cat'. Clan Sutherland's crest also bears a wildcat, sitting but with both forepaws raised, with the words 'Sans Peur' ('Without fear'), and the Duke of Sutherland's Gaelic title is *Morair Chat*, meaning 'The Great Man of the Cats'.

The Picts worshipped a wildcat goddess, Ura, who dwelt in the heather. For Pictish people, the wildcat was a creature imbued with magic – like the hare, it was linked with the moon and with witchcraft. There are many representations of wildcats to be found within old Pictish stone carvings dating back to the Iron Age, though fathoming out their cultural significance is probably impossible. Interpreting later representations of cats on artefacts is even more difficult, as domestic cats came to Britain with (and a few before) the Romans and had become very widespread in Britain by the seventeenth century, so some British and Scottish cat-related legends are more connected to them than to wildcats.

One enigmatic cat of legend is the Cat Sith (also Cat Sidhe, Cath Sith or Cait Sidhe), a Celtic fairy creature native to the Scottish Highlands, sometimes characterised as the alter ego of a witch. While not wholly malevolent, it was not a spirit being to trifle with, as it was able to steal the soul from a corpse before it could be taken by the gods and could place a curse on livestock if not adequately appeased. To prevent the former, Celts would guard a newly dead body overnight and discourage or distract the cat spirit by extinguishing fires in the room, and by playing lively wrestling and jumping games nearby (which the Cat Sith, being a cat, would want to join in with). On Samhain, the winter festival, leaving a saucer of milk out for Cat Sith would bring a blessing to your house, but if you forgot then your cows' udders would be cursed and you would have no more milk from them.

Cat Sith may be based on the Scottish wildcat, but its appearance suggests a different origin: it was large, but jet-black instead of striped, with a diamond of white fur on its chest or throat. Today, many believe that it was inspired by what are now known as Kellas cats (after a Highland village near where some sightings occurred) – very big black cats that are rumoured to dwell in the Scottish Highlands. The biggest known individual measured almost 110cm from nose-tip to tail-tip but considerably larger examples have been reported. Since the 1980s, several such cats have been caught in north-east Scotland and the biometrics and genetics of these and other museum specimens investigated. The majority were found to be hybrids between wildcats and domestic cats, rather than feral black cats. Ancient Kellas cats, however, may have been pure but melanistic (black-furred) wildcats – at least one museum Kellas cat was identified as a melanistic wildcat, and melanism is known in other wildcat populations. Several taxidermied Kellas cats are on public view at the time of writing – you can see them at the Elgin Museum, Elgin, Morayshire; the Royal Scottish Museum, Edinburgh; and at Aberdeen University in the foyer of the zoology department. A couple of specimens of curiously long-faced black mystery cats have also been found in Scotland in the 1980s and 1990s, and these were briefly dubbed 'rabbit-headed cats'. However, they too are now considered to be hybrids, with the domestic parent a Siamese or other long-faced 'oriental' breed.

Incidentally, another small mystery black cat has been noted in Transcaucasia, on the border between Europe and Asia, just north of Turkey. A writer called C. Satunin published his descriptions, from several specimens, skins and skulls he examined, in *Proceedings of the Royal Society of London* in 1904. So confident was he that this black cat was a new species to science that he named it as such, with the striking moniker *Felis daemon*. The cat was larger than a domestic pet, with a black or very dark-brown coat scattered with white hairs. However, no further evidence of its presence has been found, and it's likely this, too, was a black-furred hybrid between the local European wildcats and black domestic cats.

The evidence that wildcats once occurred naturally in the wild in Ireland is patchy. It's possible the evidence of their presence is down to imported, traded living cats and corpses rather than natural occurrence of actual wild animals. There are definite cultural links to wildcats in Ireland, though. The legend of Cat Sith was known from Ireland as well as Scotland, and several notable Celtic Irish warriors wore wildcat pelts to demonstrate their fighting spirit. Cats are also acknowledged in a variety of Irish Gaelic ancient place names.

Some folk tales from further afield explore how the cat went from wild to tame. The African story 'The Cat Who Came Indoors' tells of a female wildcat seeking a companion. Her first choice, another wildcat, is killed by a leopard, so she decides to team up with the leopard instead. But then a lion kills the leopard, so the wildcat stays with the lion instead ... until he too is beaten by a bigger enemy – an elephant. The wildcat feels sure that she has found a worthy companion at last, but then a man comes upon them in the forest, and shoots the elephant dead. So the wildcat follows the man back to his hut, and lives in the roof, hunting the mice and rats that come to raid the food stores. One day, there is a noisy altercation inside the hut. The cat looks down from the roof just in time to see the man violently thrown out from the hut, and in the doorway stands his angry wife. When the wildcat sees the wife, she knows that she has found the 'finest creature in all the jungle'. She jumps down from the roof, moves into the hut, and is still there today.

African wildcats were treasured by the ancient Egyptians, who first domesticated them. One of the most beloved figures in the Egyptian pantheon was the cat-headed goddess Bastet, or Bast, deity of joy and love, who came to prominence around 1000 BC. A temple to her – comprising a shrine and a huge statue surrounded by gardens – stood in the city Bubastis, and Egyptians travelled to the city for a week-long festival each year. Bastet inspired the famous 'cat cult' whereby cats were held in exceptional reverence – to kill one was punishable by death, and when the family cat died there was a period of mourning. Dead cats were embalmed and mummified, and taken to a cat cemetery, along with a supply of embalmed and mummified mice to sustain them on their journey through the underworld. Several vast cat cemeteries have been unearthed, though sadly their historical interest was not always recognised and many valuable specimens were destroyed.

The Egyptians did not want to share their precious cats with the world, and exporting cats was made illegal. This slowed but could not curtail the spread of domestic cats into Europe and beyond. No other nation adored cats quite to the same extent, but a few came close. For example, during the Song Dynasty of China (960–1279) there was evidence that pet cats were treated lovingly and that special foods and treats for them could be bought at market. The cat does not have a place in the Chinese zodiac, but in the Vietnamese equivalent, the rabbit is replaced with the cat (probably due to linguistic confusion).

Ancient Greek and Roman cultures were familiar with domestic cats and kept them as pets as well as pest controllers. Cats make the odd appearance in the mythology of the time – the Greeks added cat associations to their ideas about the hunting goddess Artemis after linking her with Bastet. The same went for the equivalent Roman goddess Diana. In Ovid's epic poem *Metamorphoses*, she transformed into a cat when the Roman gods had to flee to Egypt. In general, cats were looked upon quite favourably but their mischievousness and 'lecherous' character was also noted. The Islamic faith also holds cats in high regard, partly because of their fastidious ways and partly out of admiration for their beauty. There is a story that, when Muhammed's favourite cat fell asleep on the sleeve of his robe, the prophet cut off the sleeve to avoid disturbing the cat.

In Britain, domestic cats and wildcats were both present in the years following the Roman invasion (though the former was gradually increasing and the latter slowly declining). It is difficult, therefore, to be certain whether some of the cat-related myths of the time related to one, the other or both, though over time it's likely the balance shifted more and more towards domestic cats. Through the Middle Ages and into Victorian times, cats of all kinds in both Britain and western Europe were linked closely with witches and witchcraft, which made them enemies to the Church, and many thousands of them were killed (along with the alleged witches). It would seem fitting that a witch would choose a wildcat over its tame, milk-lapping cousin as her familiar, though at least some people of the time – particularly those of a highly religious persuasion – were almost as suspicious of the latter as the former, regarding both as conduits of evil. In the fifteenth century the Duke of York wrote of cats thus: 'Their falseness and malice are well known. But one thing I dare well say that if any beast hath the devil's spirit in him, without doubt it is the cat, both the wild and the tame.'

In some regions, this mass antipathy towards cats caused domestic cats to become rather scarce for a time, and this had a most unfortunate unintended consequence of facilitating the spread of bubonic plague, which began in the fourteenth century. The disease was carried in the digestive tracts of fleas living on black rats, and with fewer cats around to kill the rats, they and their fleas spread unchecked and more widely. It took some time for people to link the plague with the rats in particular, and for a while it was widely believed that any flea could spread the Black Death. Consequently, even more cats were killed, until people began to notice that those people who held on to their cats were not dying from plague at the same rate as the rest of the population. In the wake of this insight, new laws were passed in Europe that placed cats under strict protection.

Wildcats have also made frequent appearances in British literature – including several mentions in Shakespeare's works. In *The Taming of the Shrew*, the titular 'shrew' Kate is called a 'wildcat' more than once, and it's clear that this isn't a compliment. In Act 1, Scene 2, Gremio says to Petruchio, who has fallen for Kate:

> *O sir, such a life with such a wife were strange.*
> *But if you have a stomach, to't, a God's name,*
> *You shall have me assisting you in all.*
> *But will you woo this wildcat?*

Later on, Petruchio puns on the wildcat theme by saying that he will transform her from 'a wild Kate to a Kate'. If a person was called a wildcat it meant they were fierce, impulsive, ungovernable – the term survives today in phrases like 'wildcat strikes' to describe unauthorised work walk-outs. The wildcat also gets a mention in *The Merchant of Venice*, but this time in reference to its sleepiness rather than its ferocity. The 'brindled cat' that 'mews thrice' in *Macbeth* (Act 4, Scene 1, line 1) may also be a wildcat, although if it is then Shakespeare's observational powers have let him down as wildcats typically do not mew, just hiss, snarl and caterwaul.

In more recent literature wildcats tend to be more sympathetic figures, and they crop up with relative frequency in children's animal stories. As a child, I was very struck by the wildcat that appears in *Danny Fox Meets a Stranger*, one of the series of books by David Thomson that featured the wily Danny Fox and his family. The foxes live in a rough and rugged landscape, suitable for wild beasts of all kinds, and one day a wildcat comes to the foxes' den. This cat, whose name is Shaggy, explains to the foxes that they can tell she is a real wildcat by the rings on her tail. She is calm, wise and infused with magical power, which she uses to help the foxes drive away the wolves that threaten their territory. I adored her not just because she was a wildcat but because she was that extremely rare thing in 1970s children's stories: a stand-alone female character who was neither a princess nor a damsel in distress.

In the more recent, popular teen story *Angus, Thongs and Full-frontal Snogging* by Louise Rennison, the Angus of the title is the family pet but is a Scottish wildcat hybrid that the protagonist Georgia and her family brought home from a Scottish holiday. Angus looks just like a wildcat and his other wildcat traits include immense size and grumpiness, though in the 2008 film adaptation of the book he is played by an emphatically not-wild fluffy grey Persian. In the enormously popular *His Dark Materials* books by Philip Pullman, each human is accompanied by an animal 'daemon', a manifestation of their spirit which shapeshifts through different species through the early years of the human before settling on a final form at the start of the human's adolescence. The lead character, Lyra, has a daemon called Pantalaimon who, before he takes his permanent form of a pine marten, often takes the shape of a wildcat. It was one of his more useful forms – in wildcat mode he is a fierce fighter and defeats several enemies in battle.

However, wildcats are missing from many of the animal-anthropomorphising classics of children's literature. I've seen it argued that this oversight explains why we as a nation aren't as interested in wildcats as we should be – and it's certainly true that rose-tinted childhood memories are good for fostering warm fuzzy feelings towards wildlife. Conservation projects for water voles and badgers have *The Wind in the Willows* to look to; for otters there is *Tarka the Otter*, and so on. A few wildcat books do exist, of course, like *Chia the Wildcat* by Joyce Stranger, but they are not of classic status. Perhaps this should be a clarion call to authors of children's fiction who also care about Scottish wildcat conservation – what's needed is a universally loved and popular fictional wildcat hero to inspire the next generation. Or perhaps it needn't even be fictional. After all, the truth of the wildcat's life is more strange, moving and fascinating than any fairy tale.

Speyside, 2013

Twelve hours and 40 minutes. It's long enough to fly from London to Bangkok, or to Honolulu or Caracas. I could be spending this late Friday night wandering around a shiny airport, browsing guide books for Thailand (or Hawaii, or Venezuela). Instead, I'm sitting in the distinctly un-shiny Victoria coach station on one of a row of plastic chairs, sandwiched between two fellow travellers who look just as tired and apprehensive as I feel. We're waiting for the 588 coach from London to Inverness, due to depart at 10.30 p.m.

There are several ways to get from London to the Scottish Highlands. You can drive, you can take a train (by day or overnight on a sleeper) or you can fly, probably to Inverness. Then there's this way – the cheapest way, possibly the greenest way, and certainly the most gruelling way – the overnight coach. National Express do their best to make the journey as painless as possible and I have nothing but praise for them, but nearly 13 hours on a coach is what it is and I'm not anticipating a fun night. Especially as several of the other coaches seem to be running as much as three hours late and there's no sign of ours yet.

It's not so bad in the end. The staff open the doors at 10.30 p.m. as our coach pulls forwards, and a queue swiftly forms. I join it. The coach reminds me of a bull or buffalo, thanks to the long, arching wing mirrors – curved horns framing its blunt head. Luggage is loaded into the coach's capacious belly, and one by one we

climb on board to take our seats. The coach is packed –
there's one of those wiry, Craghopper-clad young
backpacker types in the seat beside me, and behind us a
bevy of Glaswegians conducting a loud, amiable but
very sweary conversation in their almost impenetrable
accents. We are surrounded by other rumbling coaches,
bound for Edinburgh, Cardiff, Aberdeen, Salisbury –
every one of them running late and impatient to get
going. Coordinating this mass of vehicles may not exactly
be air traffic control but it still seems a daunting prospect
for me. It's 10.45 p.m. when our coach lumbers out of
the station, leaving its rivals behind, and noses its way
into the central London traffic, still heavy and relentless
even this late in the day.

I'm on my way to Aviemore. First, though, we will call
at Golders Green to fill the handful of seats still vacant.
Then the long plod through the night northwards to
our first stop, Penrith. Then into Scotland, where we
stop at Lockerbie, then Glasgow (where I presume our
sweary friends will leave us). Then Stirling, then Perth,
then Pitlochry. And, by nearly lunchtime tomorrow,
we'll reach Aviemore, where I will leave the coach and
its few remaining passengers to complete their journey
to Inverness.

For now, I stare out of the window at the London
sights as they crawl slowly by. I notice a couple of
sculptures I've not seen before. In a square of green
somewhere near Hyde Park, a big chrome hand breaks
out of the earth, clasping a Vespa scooter in its shiny
fingers. The next one is the oversized, severed head of a
horse, carved from stone and balanced on the tip of its
nose. I don't know quite what to make of either of them.

As we creep northwards through increasingly anonymous London streets, the coach conductor delivers his announcement in matter-of-fact tones. *Wear your seat belt Don't drink alcohol – if you do, you'll be chucked off the coach.* (the Glaswegians, who I suspect have 'pre-loaded', chortle at that one). *Fail to get back in time from a comfort stop and we won't search for you, we'll just leave you behind. No, I'm not joking. Yes, there is a toilet on board but seriously, hang on till the service station if you possibly can because things can get very bad very quickly in there.* I huddle down in my seat and watch the rows of kebab shops and off-licences, the crowds of people going in and out of them. I imagine dense forest and wild moorland, the absence of sound. I'm on my way to wildcat country – albeit slowly and uncomfortably.

Soon we are on the motorway and there's no longer anything to look at, but it's still difficult to sleep. I'm getting uncomfy, with my camera bag on my knees (but there is no way I'm letting it out of my sight or even out of my arms). The young man beside me is dead to the world and, judging by the quietness all around, so are most of the others, including the rowdy Glaswegians. I conjure up the image of Abernethy Forest, one of my favourite places on Earth. I imagine padding down the pine needle-cushioned pathways, alone, losing myself in the stillness and the sweet air, becoming a small animal navigating a complex path between the towering rutted trunks of the old Scots pines. Raising my binoculars to scan the next ridge along, and glimpsing a broad, banded tail disappearing over the top of it.

Just as my eyes are beginning to feel heavy, we pull into our first service station. I take the chance to jump off, to stretch and wiggle and yawn, and pull my coat around me against the cold. It's only going to get colder.

Back on board, we make swift progress northwards on the M6 and I doze, dreaming of blazing green eyes and distant tawny shapes appearing and disappearing along banks of dry bracken. I don't wake up properly until we pull into the station at Penrith. We've made it to the Lake District, how about that? It's sometime in the small hours, and pitch-dark outside. The man beside me wakes, straightens, jumps up, grabs his bag and strides down to the front of the coach. He's the only one to get off here. Peering out of the window, I watch as he swings his rucksack over his shoulders and lopes away confidently into the night. With an empty seat beside me, I'm suddenly the envy of the rest of the coach – those who are awake, anyway.

It's still dark when we reach Glasgow and disgorge more than half of the passengers. Through a slow grey dawn we cross-country our way eastwards, then north. We lose a few more people at Stirling and more at Perth. Now we're on the A9, it's daylight and snowy hills are looming into view all around. Below, the Tay rushes along, a few early common gulls swirling like snowflakes along its deep-cut valley. Proper Scotland. My sleepiness is overwhelming but in the moments I'm awake my heart thrums a bit more quickly with excitement, anticipation.

At Aviemore, I stumble down from the coach, retrieve my case, and head up the high street. A strange place, Aviemore, laying low in the Cairngorms' shadow. Its name makes me think of eagles and glittering snowcaps but the town itself is uninspiring. One long straight road lined with blocky shops selling outdoor gear, coffee, outdoor gear, coffee, and so on. But when I reach Tesco I hear a familiar sound and pause. In the car park's spindly trees, there's a flock of fat little birds, peach - and ochre-coloured with outrageous punky hairstyles that whip about in the brisk breeze. They fidget and bicker and twitter like sleigh bells. Waxwings. Well worth the ten minutes it takes to set down my bags, unpack half of my camera bag to unearth the actual camera, and take some pictures.

Buses from here to Nethy Bridge aren't frequent. Or maybe they are by Highlands standards – I'm not sure. Sooner or later one turns up, anyway, and carries me away from Aviemore and into the Spey Valley proper. The roads, the railway and the villages are all tucked into this strip of flat land within one of the most mountainous parts of Scotland. We skirt the edge of the great pine forest, and weave along narrow roads through meadows, riversides, the occasional reedy lochan. Among the black crows dotted amid cows is the occasional oddity: a crow that's black and grey. A hoodie. We're close to the intergrade zone where the distributions of the carrion crow and the hooded crow meet. This, as much as anything, brings home to me how far north I am now. I'm dizzy from sleeplessness but try to cast my gaze as widely as I can – along the field edges, down the trails into the forest, everywhere where a Scottish wildcat could conceivably wander. Never mind that it's broad daylight. Hope springs eternal – it has to.

It's a 20-minute walk from the bus stop in Nethy Bridge village into the forest and the little cottage where I'll be living for the next two weeks. I trundle my wheelie case down the narrow wooded lane that undulates away alongside the River Nethy, a lively tributary of the Spey. I'm stepping slowly, keeping my eyes open. The trees get taller, the houses fewer; human sounds fade away but the chatter of the Nethy is ceaseless. My gaze is mainly sweeping far ahead down the lane, in case any animal should rush across ahead of me. But it's when I glance to my right into a pasture that I see my first wild mammal of the trip – an ash-coloured roe doe who trots hastily to the edge of the trees when she sees me and watches me go by from there. A moment later, I see my second – a red squirrel bounding across the large garden of a particularly grand house on its way to raid the purpose-built squirrel-feeder nailed halfway up a tall pine.

This all seems to augur well. I find my way to Dell Cottages and locate mine among the small group of single-storey buildings tucked into the north-eastern edge of Abernethy Forest behind a wall of pines and silver birches. I pause at the door to listen. I can't even hear the river now. Instead I can hear jackdaws, the crisp 'kik' from a passing great spotted woodpecker, the squeaks of coal tits and 'pinks' of chaffinches. More distantly, cows, pheasants, the bugling of a passing skein of greylag geese. Actual wild greylag geese, winter visitors from Iceland. Not the semi-feral herds that loiter around park lakes back home. Each new voice sparks excitement, wards off the urge to sleep. I glance at the bed, then dump my things on the living-room floor, grab my camera, and walk into the forest.

Why am I here? I know it isn't the best place I could be. There are wildcats here – or at least we think there are – but their gene pool is polluted heavily with the DNA of domestic cats because this is (by Highland standards) an area with a lot of people, and a lot of people means a lot of pet cats. Even if I were to see a cat that looked convincingly wildcat-like, the chances are it'd be a hybrid. I should be further east in the remoter Cairngorms, or over in the far west in Ardnamurchan, or north in Caithness, a region actually named for wildcats. I should be settling into a camouflage-patterned bivvy on a wild moor five hours' hike from the nearest road.

Instead, I'm wandering into this large (but still remnant) Caledonian pine forest. It looks unspoilt, untouched, but it's not, of course. There are human-made pathways, for a start – wide, straight ones like the one I'm on now, and narrow wavering ones that zigzag away into the dark latticework of the distant trees. There are plenty of gnarly old pines but there are younger ones too, stands of them, marking places where clearance happened in the not-too-distant past. I pass a clearing and note a row of telegraph poles and their cables. There are signs up here and there, pointing visitors towards the surrounding villages and warning of the hazards of fire and off-lead dogs. Not untouched, by any means, but there is real wildness here. Away from the trails the ground is rugged and pitted, carpeted with a fabulous thick understorey of heather and bilberry, dotted with boulders, great fallen trunks and the spaces between them that offer shelter and safety. It's potential wildcat habitat. If I'd come here a hundred years ago, I might have stood a fair chance of seeing one. And if I could somehow come here a hundred years hence, if current conservation efforts have gone as well as they can and all our hopes and dreams for the Scottish wildcat have come to fruition, there could be good numbers of wildcats here again. But in the present moment, it's more about just being here than any real chance of seeing what I long to see.

March is a quiet time in the forest. Summer's visiting birds aren't here yet – the likes of willow warbler, tree pipit, cuckoo and redstart. They're heading north but it'll be weeks before they arrive. That leaves the residents – the coal tits, the chaffinches, the crossbills (three different species, though you'll need a sonogram of their calls to be sure which is which). A soft purring note from the trees above alerts me to another the crested tit, found nowhere else in Britain except these forests. I stare almost vertically upwards and with some difficulty discern the bird poised momentarily on a dead branch tip – a silhouette against a pale cloudy sky, but there's no mistaking that spike-topped head profile.

I can't keep going much beyond this. I would like to lie on the cushiony heather and fall asleep but it's cold and I fear deer ticks, so I head back to my cottage. It's late afternoon now. Hunger and tiredness battle for supremacy, but to satisfy the former I'd have to walk back into the village to visit the shop, so instead I climb into bed and sleep solidly for the next 14 hours.

The next day I walk to the shop for provisions, following the bumpy riverside path. I disturb a goosander on the way – a sleek, cigar-shaped diving duck. She is resting on a rock when she spots me approaching, and slips into the rushing water, letting it carry her past me. When I am almost at the village, another river bird makes itself known – the hard clicking call of a dipper sounds past my ear and I just glimpse the bird itself as it rockets past in low straight flight. It's a round dark shape on a whirr of wings. It reminds me of a skimmed stone or a bomber plane, overcoming its own heaviness through powered straight-line speed. Then, like a thrown stone that's run out of kinetic energy, it drops into the water and the rivulets engulf it immediately. An alarming moment, but this is what dippers do. I know the bird is quite comfortable underwater, whether it's walking about clasping riverbed stones or being carried along by the flow.

I wonder what it's like to be a dipper, never to leave the riverside. To live every moment of your life with the white-noise roar of rapids in your ears – is it like silence to them? What do they hear when they submerge; what can they see? I spot the bird again as I make the return walk with a bag full of provisions. It sits on a midstream rock in a field-guide pose and preens its sheeny dark and snowy-white plumage, batting pale eyelids and bouncing on its sturdy legs. I leave it behind and follow the trail home. Time for a long walk.

The RSPB's Loch Garten Osprey Centre lies at the other end of the forest. It's not open yet (too early in the year) but the walk is a good one, covering a good 13km of forest. A lot of it is along the Speyside Way, a 96km trail that follows the great river from the Spey Bay on the Moray coast all the way to Aviemore. This particular stretch takes you away from the river and through the forest, but you'll rejoin it if you carry on south to Boat of Garten, or indeed north to Grantown-on-Spey. Its signposts bear thistle heads, and they guide me west out of the village, along a lane for a while and then into a section of forest that is new to me.

All is so quiet. It inspires me to quieten as well. I'm soft-stepping, wincing when my boot cracks a twig or my coat makes a swish. I know there is wildlife here, rare and localised wildlife too. The capercaillie, a turkey-sized grouse, is in this forest, albeit in dwindling numbers. For such a big bird it is amazingly skilled at silently slipping away well ahead of searching eyes. I've only ever seen one, a male, spotted by a birding friend as we walked a quiet trail in a patch of forest rather like this one. The bird – huge, black-feathered, utterly imposing – stood alert with its fine fan-tail half-raised on the side branch of a pine. It was dozens of metres away on the far side of a steep valley, well outside the 'scare' zone of any other bird. But it was watching us, and when it realised we were watching it too, it turned its back, stepped off its branch into the air and flew rapidly away from us into the shadows. This extreme sensitivity to disturbance could be one of the reasons why capercaillie numbers are in free fall. It would probably be best if I didn't see one, and the chances that a caper would be hanging around so close to a main walking pathway are slim anyway.

The path takes me through a patch of open, heathy ground, alongside boggy pools which, later in the year, will no doubt teem with exciting insect life. The Highlands has dragonflies that don't occur down south – white-faced darters, northern emeralds and the like. But in March the dragons are still water-bound nymphs, and the skies are empty of their skimming forms. A lone buzzard wafts over, drawing my attention with its desolate call – a soft cat-like wail that is, I suspect, the closest I'll come to a feline encounter today.

I near the visitor centre, and follow a short path towards it, and then I have a sudden, unsettling sensation that I'm being watched. I stop, and turn slowly. There is a small birch next to me, not much taller than I am, and in it is a coal tit, barely an arm's reach away. It's uncharacteristically still. It is, indeed, watching me. And so is the other one sitting close to it, and so are all five of the other coal tits perched here and there in the same tree. They tilt their pied heads up and down, side to side, and are quiet. I feel suddenly thrown into a bizarre hybrid production of Disney's *Snow White* and Hitchcock's *The Birds*. I walk on, slowly, and in each tree I pass there are masses of coal tits. A few chaffinches too, and a robin. And then there is a crested tit, impishly beautiful with its banded face and speckled spike-crest. In its front-on stance it's the shape of a teardrop as it eyes me from its perch on the arm of a bench. I realise what's going on − I'm approaching the RSPB's feeding station here and these birds are taking a break from jostling over that, instead hoping for an easy handout from me. I resolve to return with a bag of peanuts later in the week.

I turn to follow the path back through the forest, and soon I have left the unsettling army of little birds behind and am back in the quiet and still. I walk in a reverie that's almost uninterrupted by wildlife until I'm quite close to the edge of the village. There, a sudden soft whirr of wings makes me snap around just in time to see a female capercaillie vanish at speed between the trees.

The next day, drizzle is falling and I take two buses southwards to Kingussie and the Highland Wildlife Park. The staff on the gate are surprised to find a visitor arriving car-less, and I have to wait until a keeper in a safari vehicle turns up to drive me through the first section of the reserve, where big, hooved beasts wander unconfined. I'm delivered to the central part of the park, and begin to explore.

This small zoo holds mainly European – in fact mainly British – wildlife, though there are exceptions. I'm here in time to see the two young male polar bears, recent arrivals, roaming their extremely spacious hillside enclosure. I've been watching for a little while when the bears come together and begin to play. It's a charming but also intimidating spectacle, the two animals' immense power all too obvious as they stand tall and wrestle, grumbling and grunting and mock-biting at each other's faces and necks. The effect is only slightly spoiled by the fact that both have clearly been having fun in the muddy patches and their coats are grubby, making them brown enough to be *Ursus arctos* rather than *U. maritimus*.

I wander around the rest of the park. There is much to divert me both behind the wire and out in the open. A small flock of oystercatchers is flying about – I keep bumping into them as they flash overhead, piping excitably. On the distant hillsides, herds of red deer are grazing. In a tall aviary, a snowy owl rests on her perch, watching me with sleepy cat-eyes. I make my way past the Eurasian lynxes, which recline together on the roof of their shelter, unphased by the increasing rain. One lynx is sleeping on its side in an elegant muddle of long limbs, while the other is awake with head raised, gazing into the middle distance, its beautiful tufted face turned away from the handful of admiring humans at the cage wire. I wonder whether some instinct tells it that its kind once belonged here, in this wild landscape. Our lynxes were wiped out some 1,300 years ago. Now, it's looking increasingly likely that a lynx reintroduction or reintroductions will happen, sooner rather than later.

I follow the signs to the Scottish wildcats, passing red squirrels (showy) and capercaillies (hiding) along the way. There's their enclosure – or enclosures. The design is clever; the wildcats have several 'rooms' to live in, linked by tubular overhead walkways that cross the paths. I look up into the walkway over the path ahead and meet the cool stare of a big male Scottish wildcat, sitting bunched up in the mesh tube just where it bends round a corner. I stop and drink him in.

The signs on the cage tell me his name is Hamish. He is huge, even in his compact posture – a round-edged loaf of cat with his fat banded tail wrapped around his toes. His face is full, square-jawed. His tremendously thick coat looks bristly, as though it'd feel harsh under a stroking hand, not that the look on his face encourages stroking. Far from it. The close-set green eyes stare, a confident challenge. The black markings that surround them are bold brushstrokes. On his chest is a star of white, which I know or suspect should not be there. White markings on apparent wildcats can be a sign of domestic genes. A lot of white is a clue that the cat in question will not prove to be a wildcat or even a 'good' hybrid, but a touch of white is not so bad. It seems pretty feasible that a little white chest-star marking could occur in pure wildcats. When I speak to his keeper later, she says that it can and, although Hamish (like every other wildcat in captivity) doesn't have a completely pure wildcat genome, the DNA tests he's had show that he isn't far off. In every other aspect, he really looks the part, from his deep-pink nose to the blunt black tip of his tail.

The other wildcat in the enclosure doesn't match my mental picture as closely. But part of that is that she's female and some of the traits we associate with Scottish wildcats – like size and broadness of head – are more pronounced in males than females. Her name is Betty and she has a slimmer tail than Hamish. On the other hand, she does not have a white chest star, her tail is ringed and black-tipped like it should be, and her beautiful face has its own undeniable strength and breadth. I learn from the keeper that these two had kittens last year, which have recently moved on to new homes, and that Betty still misses them. Both of the cats are soon distracted by lunch – the keeper shuts off one of their rooms and hangs a fat, fluffy rabbit leg from a rope inside. Hamish heads straight for the room when it's opened up, and stands on his hind feet to grab and tug at the bounty.

I stay with the two of them for a while. I take photos of them and of the posters that highlight the plight of the Scottish wildcat and the efforts underway to save them. Captive cats like Hamish and Betty are incredibly important, especially Hamish with his exceptionally good genes. Showing them to the public is no less important, for wildcats are at our mercy – even more so than most wild animals.

The day after that, I retrace my steps back to the Osprey Centre, this time with the promised bag of peanuts. The weather has deteriorated, a few snowflakes waft about as I arrive, and the little birds are hungry. The coal tits fly to my hand to feed, just as I thought they would. I can't photograph them as they do because my lens is too long (or my arms too short), but it's not a problem – the closeness is breathtaking and the touch of their feet on my fingers sparks my soul. Standing there alone in the first moments of a snow flurry, hand-feeding wild birds in a wild place, I feel emotional, lonely, content.

A slow plod back and I'm nearly at the village again, not far from where I was when I saw the capercaillie two days ago. I'm crossing a small patch of sedgy grassland when something brown and furry bolts out of a hidden hollow through the long vegetation. I get a glimpse of a dark-striped back and a thick, black-tipped and black-ringed tail, and adrenaline sets my heart pounding as I watch the animal race to a big birch tree and start to climb. Then the excitement fades as I see that the quickly scrambling legs are white and so is the belly, and halfway up the cat stops, turns to me and shows a snowy chest and a face marked with white cheeks and blaze. Yet its eyes – intense and gold-green and sparking with fright and fight – remind me of Hamish's eyes, and it has a wildcat's thick coat and a wildcat's wide face and a wildcat's tail. I take what photos I can of it through the branches before it carries on climbing up and away.

I walk in the woods every day after that, miles and miles, varying my route, shifting my timings around, exploring open areas as well as the heart of the forest, sitting still and hidden for as long as I can bear on hillsides that afford a wide view across forest edge, marshy field, rock-strewn flank of open moor. At midday I go running, plodding along the cushiony forest trails in their cathedral calm. I see almost no other people and I see wildlife everywhere, but not the one wild thing I'm after. I find a favourite place on the forest edge, where my hiding place overlooks a bracken-covered hillside like the one I dreamed of on the coach. I go there at dawn and again at dusk. I stay late into the evening, until the light has nearly failed and my tired eyes imagine movement – the rocks and stumps coming to life.

In my second week, the wintery weather arrives in earnest. I wake up in a world muffled by snow, the trees around the cottage thronged with hungry birds, the spring flowers poking their leaf-tips disconsolately through the cold blanket. One morning I watch a red squirrel cavorting in my cottage garden for an hour, rushing and bounding around for no obvious reason, marking the pristine snow layer with a veritable art installation of spidery footprints. I fill and hang a bird-feeder from a tree in front of the garden door, and am rewarded with non-stop visitors – one day, a pair of fat-billed Scottish crossbills (well, probably) come and sit in the tree, and the next everything is kept away by a beautiful, wild-eyed young male sparrowhawk, lurking not nearly as invisibly as he thinks he is.

My wildcat searches carry on through the snowfall and the thaw. I do see one more apparently feral cat later in my stay, hunting in a swampy meadow well away from the village. This one, dark and brindled all over, gets my heart pounding fast, but a look through the telephoto camera lens shows that she is not tabby but a dark tortoiseshell, and while she doesn't exactly approach me for a cuddle when she notices I am there, she also doesn't show anything like the alarm I'd expect from any cat with some *grampia* in its genetic make-up – she simply

carries on staring into the grass. A day or two before I
head home, I talk to a local woman and she tells me of
the litter of blunt-tailed, stripy kittens her childhood pet
cat produced one year. They were just something
interesting to talk about back then, these part-wildcat
kittens. Now she knows that hybridisation is leading the
Scottish wildcat to extinction, and she has no cat of her
own anyway. But many others don't know, or don't care,
and allow their unneutered pets to wander at will. Any
true wildcats that remain around here are far more likely
to meet a feral or pet cat than a fellow wildcat when it's
time to mate.

I feel melancholy on the coach ride home. Leaving
this beautiful place brings sadness, and although I had
never really expected to meet a wildcat in the wild, the
scale of their absence is more obvious to me now than
before. Yet walking in their world was also uplifting.
I fall asleep somewhere on the M6, somewhere after
1 a.m., and dream that there had been time for one last
slow, silent walk through the trees, one last pause to
breathe and gaze, and one last heart-shaking moment
of hope.

CHAPTER THREE

Born to Kill

Sometimes my former-feral cat sits up by the window and watches birds outside. As she watches, her teeth chatter and she makes a series of staccato squeaks. It's hard to tell through the fluff but I know her body is tense, her muscles bunched in excitement. Every instinct fixes her focus on the quarry. She is, actually, not a capable killer – I've known her to take down a housefly but nothing else. Had she grown up fully feral she'd probably have had to scavenge to survive, or more likely she'd have starved. For pet cats, being able to hunt is optional. For feral cats, it's extremely useful, though many can get by on stolen scraps and handouts. But for

Scottish wildcats, which avoid humans as much as possible, hunting is the only way to live.

Cats of all kinds are obligate carnivores – they have to eat meat to fulfil their nutritional needs and most will eat nothing else ever. Accordingly, they are highly specialised to the task of stalking, chasing, catching and killing vertebrate prey, almost invariably as a solo endeavour. This is written into every detail of their anatomy and behaviour. The traits that make domestic cats such lovely pets – their pliant cuddly bodies, soft fur, cushioned paws, playfulness, tendency to nuzzle and their celebrated self-reliance – all stem from the way they make their living. Perhaps the most natural comparison to draw is with our other favourite pet, the dog, which is also a carnivore (albeit not obligate – dogs can manage on an omnivorous diet) but takes a very different anatomical and behavioural approach to its work. Dogs are, of course, much more variable in their anatomy than cats, thanks to the centuries of selective breeding that has spawned so many breeds for different tasks. So for the purpose of this exercise, the dog we have in mind is a fairly generic one, probably close in appearance to a wild wolf.

To start with the bare bones, the differences between a cat's and a dog's skeleton are quite apparent, though the two have the same basic spinal 'formula': eight cervical (neck) vertebrae, 13 thoracic (upper body), seven lumbar (lower body), three sacral (pelvic), and up to 23 caudal (tail) vertebrae. However, the way those vertebrae join together and to the skull is different. The discs between the vertebrae are much more springy in cats, allowing the body to bunch up and elongate to a much greater degree when the cat is running and leaping. The pelvis and shoulders are also anchored more loosely (by tendons and ligaments) in the cat than the dog. The cat's stride is, therefore, relatively longer than the dog's and it can accelerate much more quickly. Cats are also much less likely to suffer the degenerative back trouble to which dogs are prone.

The cat also lacks a nuchal ligament. This ligament connects the second cervical vertebra to the first thoracic vertebra in dogs (and also in grazing mammals, and humans). It helps to support and stabilise the head while the dog is running, especially when running with its head low. In cats, the stabilisation is done with strong neck muscles; in dogs the neck musculature is proportionately much less strong.

The cat's limb bones are proportionately shorter than the dog's, and their associated muscles are proportionately bigger. These differences aren't that obvious on the intact and furry animals themselves, but under the skin they are clear, and mark out the different ways that cats and dogs run. Picture the cat as a 100m sprinter – muscular and stocky. By comparison the dog is a long-distance runner – thin and gangly. One's all about power, the other about efficiency. The sprinter will burst out of the blocks much more quickly than the long-distance runner, but will run out of energy much sooner.

The make-up of the muscles also reflects this different approach to running. In cats, most of the skeletal muscle is made of 'fast-twitch' fibres, which can function anaerobically for a short time (*i.e.* not using inhaled oxygen to drive the biochemical reaction that generates energy, but achieving the same goal using just the stored glucose in the body). Anaerobic metabolism delivers energy very quickly but is inefficient and can't be sustained very long as it causes lactic acid to form in the muscles, which results in muscular fatigue and pain. Dogs' muscles, though, contain a more equal mixture of fast-twitch and slow-twitch muscle fibres. Slow-twitch fibres can work to full efficiency aerobically, using inhaled oxygen as fuel, so using them is much more efficient and sustainable.

This is why dogs, and their wild relatives, run their victim to exhaustion. A wolf pack pursuing a caribou may chase their quarry for hours on end, running steadily after it and waiting for it to weaken. They almost never give up – it's clear why we use the word 'dogged' to describe single-minded persistence. Cats, though, go for the surprise dash-and-pounce – if the strike misses, they have not committed much time or energy and can try again fairly soon afterwards. However, most cats' maximum speed is not exceptionally high. While the cheetah can briefly hit a top speed that beats any other land animal when racing after prey, an average household cat's fastest pace can't approach that of an average household dog. In the familiar dog-chasing-cat scenario, if both start running at the same time with the cat in the lead, the cat should escape every time, but only if it can reach safety within a few strides.

The best safe place is often 'up a tree'. But what goes up sometimes comes down, especially when you are panicking a little. If a wildcat falls from high in a tree, or a pet cat falls from a high window, its anatomy and reflexive behaviour provides it with a safety net. The athletic spring in the back, legs and neck that makes the cat such an impressive sprinter is also part of what allows it to right itself when falling and to survive the drop. If a cat falls upside-down, it quickly turns its neck and the spine begins to twist too. Turning right side up means the cat will land on its feet and thus have a much better chance to safely absorb the impact of landing because the legs can 'give' when they hit. The looseness of the feline body skin is the other part of the trick. It helps to decelerate the cat's fall, effectively turning the body into a parachute that makes use of air resistance to slow down the cat and increase the possibility of it surviving a terminal velocity fall.

Returning to our two skeletons, you'll notice that the eye sockets or orbits are a lot larger in the cat's skull than the dog's, which is to accommodate proportionately much bigger eyes. Cats are typically more nocturnal than dogs and so need big eyes for better low-light vision (more about that later). The cat's skull is also shorter, wider and more robust than the dog's, giving a stronger bite force. The dog's longer snout is better at getting an initial grip – dogs prefer to grab and shake or (doggedly) hang on to a bit of their prey instead of going straight for a lethal bite. A longer nose also means a better sense of smell as there is more surface area for the olfactory mucosa – the membrane inside the nose, which contains the sensory cells that trap odour particles. And it means a bigger mouth too when the jaws are open – useful when the dog needs to keep cool through panting during those long chases.

Cats (domestic and wild alike) all have the same dental formula, which is expressed as I3/3 C1/1 P3/2 M1/1 (i.e. on each side of the jaw, counting back from the centre, there are three incisors on the top and three on the bottom, one canine top and bottom, three premolars on top and two on the bottom, and one molar top and bottom). Have a look in a cat's mouth when you get the chance, and you'll see at the front a row of six tiny, pointed incisor teeth on both upper and lower jaws. These rows are flanked on each side by a much bigger and longer canine tooth or fang – the canines are particularly big on the upper jaw. There is also a gap between the incisors and the canines on the upper jaw that's not there on the lower jaw. This means that the upper canines are further apart than the lower ones, so when the mouth closes they fit on the outside of the lower canines. Often the tips of the upper canines can still be seen protruding when the mouth is closed. These big teeth are used to inflict the deep puncture wounds that usually end up killing the cat's victim.

Seeing the premolars and molars is more difficult as they sit further back in the mouth. Unlike our own molar-type teeth, which are flat-topped and used for grinding, those of cats have pointed cusps and are used for shearing flesh from bones, with the upper and lower teeth working together. The extra pair of premolars in the upper jaw sits a short way behind the canines – these teeth are tiny with a single cusp, but the premolars are otherwise large with multiple cusps, and the molars are also stout and pointed. The rear upper premolars and front lower molars are especially big and are also known as carnassial teeth.

Dogs and most other carnivores have more back teeth than cats do – the canine dental formula is I3/I3, C1/1, P4/P4, M2/M3. The teeth are also larger overall and a bit more uniform in size and shape, reflecting a more generalist diet.

When the paw of a cat is in a relaxed and natural position, the claws are fully withdrawn. Compare your cat and dog skeleton and (provided they've been put back together properly) you'll see the dog's claws are in contact with the ground but the cat's are not. Cats' claws are therefore not subjected to much everyday wear. They also don't wear away at the tips alone because the keratin that forms the claw grows in layers that encompass the whole claw and then loosen as a single piece. So when the cat scratches a tree or other rough surface, the entire loose outer layers of the forepaw claws can be sheared away, leaving even sharper new points below. You might find these keratin sheaths embedded in your cat's scratching post or in the arm of your sofa. You may also see your pet cat loosening and removing old claw layers during grooming by biting and tugging at the claws of both front and hind paws with the incisor teeth.

Sharp forepaw claws are important to cats for many reasons. They provide grip for climbing, and traction for fast running and turning. They are used to grab and hold prey, and they are effective combat weapons when the cat has to fight. The hind-paw claws are never as sharp as the forepaw claws, but they are used in different ways – in particular, for kicking powerfully at a victim that's already gripped in the forepaws.

Both cats and dogs have reduced clavicles (collarbones) and the clavicle is not fixed to the scapula like it is in humans, but is 'floating' – held in place by muscle alone. This gives the front limbs much freer movement, and means that the cat can squeeze its whole body through any space that its head will fit through (though this ceases to be true for some very overweight pet cats).

The feline senses are also optimised to suit the specific way the animal hunts, from finding prey in the first place to chasing, catching and killing it. Cats' eyes are adapted to a certain way of seeing, and it's not the same as the way we see. I remember being told when I was a child that cats lacked colour vision and didn't have as good visual acuity as humans did. It had seemed to me that most of what I learned about animals placed them at a considerable advantage over humans – they were faster, stronger, with better senses … in fact were better at nearly everything. So I was surprised to discover that those big beautiful feline eyes were so much worse than my own somewhat shoddy pair, and could only see a fuzzy monochrome version of the world. Of course, it turned out that what I was told wasn't quite accurate. Besides, 'worse' and 'better' are relative terms – it all depends what you need your eyes to do for you. In eye terms, dogs and cats are much more similar to each other than either is to humans.

It is true that cats' colour vision is pretty poor when compared to ours. The cells in the retina that detect colour in bright light are called cone cells. We have three different kinds of them, but cats lack the kind that are sensitive to the longer wavelengths of visual light, *i.e.* red hues, and nearly all of the cone cells they do have are of one type – the kind sensitive to yellowish-green light. Also, cats have far fewer cone cells than we do – about 28,000 per mm^2 at most, whereas we have up to 180,000 per mm^2. So cats see fewer colours than us. Most biologists believe that they see the world in muted pastel tones, but without any red tones at all, and it's the same for dogs. Us humans, with our fruit-eating primate heritage, have inherited a much more colourful world than our favourite pets have.

But the retina contains another light-detecting cell type. These cells, the rods, are sensitive not to colour but to white light, and cats have up to 463,000 of them per mm^2 compared to our maximum of about 150,000 per mm^2. Crunching those numbers shows us that humans have about four rod cells for every cone cell on average, while cats have about 25 rods per cone. Cats, therefore, can see more clearly in very low-light conditions than we can and are much more sensitive to the small contrast changes that mean an object is moving across the visual field. They also have more sensitive peripheral vision than we do.

Mammals that are mostly nocturnal tend to show an array of eye features adaptive to better low-light vision. The first is, as described above, a preponderance of rod cells in the retina. The second is having big eyes, giving a large light-gathering opening on the outside (the pupil) and a large light-detecting surface on the inside (the retina). Cats have remarkably big eyes for their face size; proportionately sized eyes on a human face would be about double the size. The cat iris – the coloured part of the eye – takes up most of the hemisphere of eye that is exposed to the air. It is a sphincter muscle – a muscle with a central hole that can open wider or tighten and constrict as required, in order to control the amount of light that enters through the pupil. The iris of smaller cat species has a figure-of-eight shape with an elliptical rather than round opening, so the pupil is slit-shaped when the iris is fully constricted. The fully dilated pupil is almost circular and very large, leaving only the narrowest ring of iris visible at its edge. The eye in this state has enormous light-gathering potential.

There's another feline light-catching trick hidden in the eyes. Take a flash photo of a cat at night, with its pupils in a fully dilated state, and you'll get that classic eerie green glow – the eyeshine that inspired the English businessman Percy Shaw to invent the reflective road stud or 'cat's eye' in 1934. A flash photo of a human's eyes at night, by contrast, will just show a pair of red pupils, as the flash lights up all the blood vessels at the back of the eye. The eyeshine is a reflection of light from a membrane called the tapetum lucidum that lies behind the retina. Light that hits this reflective 'mirror' membrane is bounced back into the retina, giving the rod and cone cells a second opportunity to be triggered by the light. This can mean that visual acuity is reduced in very strong bright light as there's simply too much light reaching the retina, but cats tend to avoid brightly lit places and in any case are usually fast asleep at the brightest times of day. Dogs and many other animals have a tapetum lucidum too.

Wildcats and tabby-patterned domestic cats have an area of pale fur immediately around the eye, most noticeably on the lower and inner edge. This is thought to be related to social signalling, but it may also have a visual function, reflecting extra light into the eyes. It is a trait common to almost all species of wildcats; even those with relatively unmarked coats, such as lions, pumas and caracals, still have the creamy-coloured 'spectacle' markings. Many other mammals, including wolves and some other wild dogs, have similar facial markings.

The shape of the cat's eyeball, optimised for maximum light-gathering through its large pupil and retina, also provides a big visual field – about 200 degrees, compared to about 180 for us. However, these advantages come with a trade-off in terms of visual acuity. Cats are short-sighted compared to humans: an object that a person with normal vision can see sharply at 60m away would not be crisply visible to a cat until it was just 6m away. However, they are also more long-sighted than us and have reduced visual acuity at close range. A less developed musculature controlling the lens within the eyeball means that they cannot change the shape of the lens as much as we can, and so are unable to focus sharply on very close objects.

The cat's field of view may beat a human's, but it doesn't approach that of certain prey animals, such as the rabbit (a favourite quarry of Scottish wildcats), which has a visual field of nearly 360 degrees thanks to its eyes being set on the sides of the head rather than at the front. The rabbit thus gives itself the best possible chance of seeing the cat coming. There is virtually no overlap between the visual fields of its left and right eye, which means that its depth perception is not especially good, but for a prey animal, it's more important to spot danger from all directions than to accurately judge how close or far the danger is – best to just leg it in whichever direction looks the best bet to keep you safe as soon as you feel under threat.

The cat's narrower visual field, by contrast, has considerable overlap between the left eye's view and the right. This means the cat can instinctively compare the position of the object of interest in the two overlapping visual fields and triangulate its exact distance away. This is a vital skill for a predator that relies on stalking to get close, sprinting to get within reach, and finally and most importantly making a pinpoint accurate pounce on its target. When a wildcat is trying to catch a rabbit, the two animals have completely opposite visual goals. The wildcat's aim is to focus on and follow the rabbit as accurately as it can and to the exclusion of everything else, while the rabbit's goal is to see as many avenues for escape as it can and constantly adjust its path as it flees.

Both the wildcat and the rabbit have eyes that work well in the dark. Seeing fantastically well in minimal light conditions is a talent shared by many other nocturnal mammals. Perfect colour vision isn't much use in the dark – it's far more useful to be able to detect the smallest movement – and this goes for whether you are a hunter or more likely to be hunted. Wildcats are potentially both predator and prey, so they need to be aware of other animals moving around them at all times.

For a Scottish wildcat, on a starless and moonless night on the moor or in the forest, highly sensitive, low-light vision is extremely important, but the other senses also make vital contributions to building the cat's sensory world and giving it the best chance to find what it needs: food, shelter, a safe path, a hiding place, and another wildcat with which to mate.

Cats have highly sensitive hearing, and are able to pick up very high-frequency sounds – up to 100kHz, which is within the ultrasonic range and well exceeds the upper limit for humans (some 20kHz) and even dogs (which max out at about 40kHz). The highly mobile outer ear can swivel towards a sound source, and move independently of its partner. Most of us with pet cats will have experienced the single-ear acknowledgement, whereby if you speak to an almost-asleep cat, it shows no reaction beyond inclining one ear in your general direction. The outer ears are funnel-shaped, the better to capture sound waves.

The feline sense of smell doesn't compare to that of dogs, particularly scent-hound breeds, which possess almost-supernatural abilities to detect and distinguish odour molecules at very low concentrations. However, cats are much better sniffers than humans are. They also have a special smell-related body part that we lack – the vomeronasal organ, which is located in the roof of the mouth. In humans, this organ begins to form but the process is arrested early in embryonic development, whereas it is fully developed and functional in cats and many other mammals as well as some other vertebrates. It and its associated physiological apparatus can almost be thought of as an extra nose. You may notice that when your cat sniffs something particularly interesting, it opens its mouth for a moment and inhales while pulling a curious, almost sneering face with the upper lip slightly drawn back. This is the so-called flehmen response and is done to draw more odour molecules in to the vomeronasal organ. In wildcats, the flehmen response is most likely to be observed when the cat is checking out scent-markings left by other wildcats or getting to know a potential mate.

Scent signals are important in cat communication. They use urine and faeces to mark territory boundaries, and also apply scent from facial glands by rubbing their faces on objects. When your cat gives you a loving nuzzle, it's doing the feline equivalent of spray-painting a graffiti tag on your face: 'this object belongs to me'. The next cat to sniff your face will get the message, and may try to replace the first cat's scent with its own. The olfactory sense isn't just about communication, of course. They can use smell to locate food that otherwise doesn't draw attention, for example birds' nests on the ground.

Cats' taste buds are, in general, not as varied or refined as ours, and they probably lack the ability to detect sweet tastes at all. The natural cat diet is entirely free of sweet and starchy substances so there is no need to sense those flavours. However, cats can also taste some compounds that we cannot, such as ATP (adenosine triphosphate), the substance which animal cells break down to generate energy.

The sense of touch is important too, particularly when the cat is dealing with potentially dangerous objects that are too close for its focal range. There is a persistent myth that cats' cheek whiskers are used as a measuring device, enabling it to decide whether or not its body will fit through a particular gap or not. This isn't the case – as we've seen already, a cat can squeeze its body through any space that its head will fit through and cats can and do get their heads stuck in objects occasionally when trying to access food. Whiskers are, however, very important sensory structures. Essentially they are thickened and very deep-rooted hairs that taper to extremely slender tips. The slightest vibration will have an effect on those fine whisker tips and provide information about the cat's surroundings; even the slight changes in airflow created as the breeze moves around nearby objects can be enough to create a whisker quiver. Cats have whiskers around their mouths but also above their eyes and on the sides of their faces. The guard hairs in their fur are also a little longer and stiffer than the surrounding, insulating hairs, and provide sensory information. Additionally, as they are not long-distance runners, cats have relatively soft and sensitive paws, and readily use them to gently touch and explore objects, with claws retracted. They also lick and bite objects of interest in an exploratory, sensory way.

In short, the sensory equipment of a cat has developed specifically in order to support a particular style of hunting based on stealth and then a short, quick attack. Being sprinters rather than marathon runners also means that their bodies are very powerful for their size. Almost all species of cats hunt alone, and their hunting methods and physical strength mean that many can tackle relatively large quarry successfully (though they readily take smaller prey as well). A full-grown rabbit is not much smaller than a cat, and by no means helpless with its forceful hind-leg kick, but many domestic cats will still successfully hunt rabbits, and a few can even tackle hares.

Like most mammals living in climates with seasonal temperature shifts, wildcats have an annual moult cycle. In November, after breeding, a thicker and longer winter coat grows in, which also usually has a somewhat greyer appearance that provides better camouflage for the less intense landscape colours of the colder months. The moult to a shorter and browner summer coat begins in April the following year.

One interesting trait in which Scottish wildcats diverge from domestic cats is in the length of their gut. Although wildcats are bigger, their guts are quite markedly shorter (averaging 124cm, compared to just over 200cm in the domestic cat). This probably can be attributed to the more strictly raw meat-based diet of the wildcat because meat is highly digestible, while the more varied and less natural diet that domestic cats tend to be fed takes more work to process (many commercial cat foods contain unsuitable ingredients, such as grain). There are also differences in skull measurements between wildcats and domestics, though the growing anatomical diversity of domestic breeds is diminishing this. The same goes for the lower jaw, which is usually larger in wildcats.

These few traits aside, there is very little difference between the anatomy and physiology of Scottish wildcats and domestic cats, even though the former is a form of the European wildcat and the latter is derived from the African wildcat. European and African wildcats are usually considered different species these days but the differences are only slight, and they are blurred and confused by the increasingly great variety within the domestic cat gene pool. It's no longer even true to say that Scottish wildcats are bigger than domestic cats, as various breeds of domestic cats have been developed that outsize any wildcat. The Maine Coon is a familiar example, with some male examples of this long-haired breed weighing in at more than 8kg, whereas male Scottish wildcats rarely top 7kg. Distinguishing a 'real' wildcat from a feral or roaming pet tabby in Scottish wildcat country has always been difficult and, now that very few if any pure Scottish wildcats survive, it's usually considered impossible by observation alone.

Wildcats are 'supposed' to be more robust and broad of skull, with smaller ears and longer, thicker coats than domestic tabbies, but this is a subjective assessment. The coat pattern on a Scottish wildcat needs to be mackerel-striped brown tabby, but a vast number of pet tabbies have this pattern and colour. The 'best' physical feature that distinguishes a Scottish wildcat and that can be assessed objectively is the shape and pattern of the tail. A wildcat's tail is broad and blunt-tipped, while the tabby's tail is narrow and tapers to a point. The wildcat's tail is marked with a series of black rings that stand separate and distinct. The bands on a pet tabby's tail are connected along the top of the tail by a continuous black line. There are also differences in the body stripes: they are continuous in wildcats, but often broken and with some spotting in domestic-types.

But what does it mean if you find a cat with a classic wildcat tail and body stripes, but the cat itself is a skinny, tight-furred and long-faced tabby? Even if the tail is appended to a suitably broad and stocky body, that cat (though it certainly looks the part) may still be mostly domestic pet by ancestry, given the nature of genetic inheritance of different traits. Only an examination of its DNA will show how close any putative wildcat is to genetic purity, and many an authentic-looking candidate for true Scottish wildcat-hood has failed this test.

There is also a growing trend to introduce genes from other wildcat species into domestic cat lineages to create new domestic breeds with an exotic appearance. Best known of these breeds is the Bengal, a well-established breed that is the result of hybridising domestic cats with Asian leopard cats to produce animals with a striking rosetted coat pattern that makes them resemble miniature leopards. A more recent new breed is the Savannah, a combination of domestic cat and serval. This is a large and long-legged breed with oversized ears and the serval's friendly, exuberant personality. There is also the Chausie cat, a combination of domestic cat and jungle cat, and the Safari cat, a hybrid between the South American species Geoffroy's cat and the domestic cat. The ethics of creating these hybrid breeds is contentious, to say the least. Differences in chromosome count and gestation period, for example, mean that losses in utero and early kittenhood can be very high. And the hybrid cats are not always the easiest of pets – Bengals, for instance, are notorious for behaving aggressively to other domestic cats and many end up in rescue, their owners beguiled by their leopard-like appearance but unable to cope with the feisty reality.

The consequences of these hybrids coming into contact with Scottish wildcats are unpredictable but there will certainly be no benefit to the latter. At the moment, there are no existing or developing domestic cat breeds that aim to mimic the appearance of actual Scottish wildcats, but this would probably not be a difficult endeavour with a few generations of careful, selective breeding. A new domestic cat breed called the Toyger, with bold tiger-like black stripes on a golden coat, has been in development through selective breeding since the 1980s and already the existing Toyger cats have a remarkably tigerish appearance (without the need for any cross-species hybridisation!).

What about personality differences? Legend has it that Scottish wildcats are not only untameable but are far more fierce than any domestic cat could ever be, and are absolutely prepared to battle to the death when cornered – perhaps even when not. Stories abound of them killing hunting dogs, sometimes two or more at a time, and grievously wounding human hunters. But anyone who has worked in cat rescue will know that some feral cats are seriously capable of defending themselves as well, and it is even more difficult to make a meaningful objective assessment of a cat's willingness and ability to fight than it is to do so about variable physical traits like size and build.

In chapter 4, we look at how wildcats behave in the wild, and the important ways in which they differ from their domestic cousins.

Ardnamurchan, 2013

You can tell when you're reaching the north-western sector of Scotland because the crows change. Staring out of the car window, I have already observed (as we traversed the eastern edge of the Lake District) that there are now oystercatchers instead of magpies sitting on the fenceposts, and now the crows are changing. I am not seeing jet-black carrion crows any more but hooded crows in their smart two-tone black and taupe-grey outfits. They seem in other respects just like their southern cousins, strutting across hillsides among half-grown lambs or wafting in pairs on a breeze we can't feel. It's high summer and the hills are alive with the sounds of corvids and sheep.

We are heading along the A82, still dazzled by a winding drive through the high hills that culminated in the towering, rugged beauty of Glencoe. Now we are approaching the Corran ferry-crossing point that will take us across a tight isthmus of Loch Linnhe and from there onto the Ardnamurchan peninsula.

My excitement at visiting Ardnamurchan is considerable. My worn-out copy of *Where to Watch Mammals in Britain and Ireland* speaks of this peninsula in reverent tones. Its rugged, untouched hills and shores, its utter remoteness, provides a home for the sort of rich and multi layered wildlife community that once ranged the whole Highlands, and a last refuge for true Scottish wildcats. I'd looked at its position on maps of north-west Scotland any number of times. It often took me a while to locate the outline of this stubby finger of land prodding towards the island of Coll, as if its very existence was somehow elusive.

We reach the crossing and join a line of cars waiting for the wide-bottomed ro-ro ferry to chug across the Corran Narrows, a 500m gap of calm water. I jump out of the car and walk to the shore to watch a couple of black guillemots that are paddling in the shallows – a new bird for the trip and a favourite of mine. They are sleek and compact, their body plumage like black velvet fur, beaded with water from their recent ungainly landing on the surface. Another flies in, showing off pied wings and dangling crimson feet. They pipe softly like songbirds as they circle each other on the surface, jousting with dagger bills. It's late afternoon. Hours and hours of midsummer daylight remain for us to complete this last and most slow-paced leg of our long journey.

We roll off the ferry and into the tiny village of Ardgour, at the south-eastern edge of the Morvern peninsula. From here, it's tiny roads skimming shorelines and weaving along deep valleys as we press on westwards, through Morvern and onto Ardnamurchan itself, through ancient oak forest and out onto an endless vista of grazing moor, where I begin to scan near and distant hillsides, walls, copses and boulder piles for wildlife, for wildcats.

Serious wildcat-seekers come here and drive at a snail's pace along these little roads through the night, sweeping a bright spotlight across the open countryside in the hope of picking up an answering pair of green eyebeams. Spotlighting like this, in remote areas like this, is probably the best possible chance any ordinary wildlife-watcher has of seeing a Scottish wildcat. But it's not without its problems and it's not a route I want to take. My plan is to park the car here and there, in spots that afford a long open view, and watch and wait through the growing dusk. I know already that my chances are extraordinarily slim. But, *grampia* aside, the promised wildlife is here in abundance.

Herds of red deer are out on the hills, moving together across the grassy moorland, grazing as they go. They are fiery, burnished in the light of this long sunlit evening. The sexes stay largely apart at this time of year, so we're seeing parties of hinds with their half-grown, still faintly spotted calves hanging close to their heels. Elsewhere there are stags, sticking together in smaller groups. Their antlers are well grown now but still in velvet. In two months' time that soft skin will be torn away to reveal the bladed weapons beneath, and hormonal surges will tear apart these cordial gatherings as the rut begins. But for now all is peaceful.

The moorland road carries us past a steep bank of earth and suddenly there are little darting shapes in the air everywhere. We pull up and get out to watch the sand martins for a moment. They are a community too, but they are frantic, a stark contrast to the slowly wandering herds of deer. The chicks tucked into each burrow in the bank are close to fledging age and are endlessly hungry. Their mothers and fathers race to meet appetites that must seem to grow by the hour. They are, I guess, flying off to streams and lochans, wherever mosquitoes and midges are blooming out of the water.

There are other little birds here on the moor that are similarly busy running around to get food for the kids. A male wheatear sits on a nearby drystone wall, twitching in agitation. His bill is packed with crane flies for chicks hidden in some rocky hollow nearby, but he won't go to them until we've moved on. His plumage, which would have been pristine in coal, ash and peach a couple of months ago, is now careworn, frayed and fading, his beauty a sacrifice for the overriding necessity to reproduce. From all directions we can hear the 'weet-weet-weet' of meadow pipits, and I wonder how many of them are nurturing a literal cuckoo in their nests. Meadow pipits are declining across most of the uplands, and the cuckoos that depend on them for foster-parent services are declining too.

We skirt above a length of seashore, noticing a pretty beach busy not with people but a herd of multicoloured cattle, apparently browsing seaweed. Next, back inland to follow the southern edge of a pine-fringed lochan. Then the road carries us downhill towards our final destination – the village of Kilchoan, close to the peninsula's south-western tip. We head for the campsite, and unpack and settle into the static caravan that will be home for the next week. Our view stretches downhill to a narrow beach of rock and gravel, a view across a calm strait to Mull. Later we drive back out to the road we'd travelled before, park up at a high vantage point, and scan the slopes all around with binoculars until it's too dark to see.

I walk out early the next morning, following the road west and steeply uphill to the spot where a derelict house stands, marking the end of the road. The road is replaced by a wavering trail through long dewy grass and over the occasional rivulet, onto the hulking face of the headland where things get too steep for me to continue. I sit on a boulder and look out to sea, watching gannets drift westwards in twos and threes. They are made tiny by distance, their spear-shaped white bodies carried on stiff black-tipped wings a few metres above the waves. Early sunshine gives way to growing cloud cover. A male yellowhammer glows upon the remains of a wall – periodically he tilts back his head and blasts out the rattle of Morse-code syllables that passes for his song. I glance behind me and notice a much less colourful bird sitting on a rock, looking plump and phlegmatic. Ash-grey and wide-eyed, it's a juvenile wheatear, days out of the nest and too young to feel anything but curiosity towards me. I turn awkwardly to take its photo – a rare opportunity. In south-east England, breeding wheatears are few and far between. They come through on their migration in autumn, but by then the young ones like this will have replaced their smoky-grey plumage with a new first-winter coat of ginger, brown and cream.

Later we drive north, towards Sanna Bay. Halfway there, we stop on the roadside to point our binoculars at a huge bird against the grey-white sky, gliding on flat wings in the distance and, regrettably, getting more distant. Hooded crows swarm around it, looking bluebottle-sized alongside the vast sweep of its wings. White-tailed eagles were persecuted to extinction in Britain in the early twentieth century and reintroduced to north-west Scotland in the 1970s. These birds are slow to mature and each pair only produces a couple of chicks each year, but they have found that this environment suits their needs and today more than a hundred pairs are breeding here. The one we've seen is a youngster, with a dark head and tail. Adults become frosty at both ends, but at all ages the bird's shape is unmistakeable, its wide square-cut wings giving it the famous 'flying barn-door' outline, with a short wedge-shaped tail at one end and a big, blocky, huge-billed head at the other.

Reintroduction programmes have been wonderfully successful at returning native species – particularly larger, predatory species – to our countryside. In England, the return of the red kite to the Chilterns and many other rural areas has gone remarkably well. Thanks to reintroduction and strong protection of the remaining native population in Wales, the bird's UK population is now storming towards 2,000 pairs, from a low of fewer than ten pairs in the 1940s. Also on the way back are corncrakes, cranes and great bustards in England, and in Wales the pine marten is being reintroduced, with the first wild Welsh-born kit recorded in 2017. If, someday, wildcats are reintroduced to Britain, it could happen here in Ardnamurchan. But wildcat reintroduction carries its own unique set of problems.

The white-sand beach at Sanna is stop-and-stare beautiful, and there's not another human here. We pick a path through marram grass-fringed dunes, alongside a shallow stream making its way to the sea. The view opens up properly – a wide bay flanked with rising banks of rocks, high tide and a pristine summer-plumaged great northern diver bobbing in the shallow water. It looks like a hand-painted wooden model, so precision-cut are its black-and-white stripes and spots, so velvet-sleek is its plumage.

I find a place to sit on the eastern side, and am joined by a flock of about 20 sanderlings, which fly down to

the water's edge in a tight little bunch. It is still only June
and these diminutive waders are in their rusty-tinted
breeding plumage, but they are not on their breeding
grounds – that's the high Arctic. I wonder if I'm looking
at birds that didn't breed at all or that tried to but were
unsuccessful. Or perhaps it's just late enough in the year
that this is a flock of adult females that have left their
chicks in the care of the fathers and are already heading
south on a migration that will take them to the southern
UK at least, and perhaps further – southern Europe, the
coasts of Africa.

I'm used to seeing sanderlings in winter when they
are white with black wing-bends and a wash of silver on
their upper sides. These are darker, marbled with grey
and black and with reddish faces – in some the red
washes further, across the chest and back. No two look
alike. But they have the unmistakeable hunched, round
sanderling shape, and the distinctive gait, rattling across
the wet sand on stout black clockwork legs. The flock is
moving into the breaking wavelets in a hesitant back-
and-forth surge, a few of the birds jumping up to fly
forwards, to get deeper. The leader of the group, up to its
tummy, begins to bathe with jerky wing-flutters, and
soon the rest are following suit. Once all have bathed,
they skitter back onto dry land and, almost simultaneously,
turn their heads to tuck their bills into their backs
to sleep.

It's a five-minute power nap. They wake up one by
one and fly up the beach, close to where I sit at the edge
of the strandline. Accepting me fully as part of the beach
furniture, they are soon trotting past me almost within
touching distance, picking and turning the dry black

coils of seaweed and snapping at the tiny invertebrates disturbed by their investigations. Two of the birds in the flock are sporting a set of bright, primary-coloured leg rings and flags. Someone, somewhere, has caught these birds and placed these markers on them with a view to tracking their migratory journeys. I take photos of both of them – when I'm back home I'll find out whose colour-ringing project this is and, hopefully, find out the history of these two particular sanderlings.

That evening, it's another wildcat stake-out, this time parked on the edge of a block of forest. As the light fades and colours diminish to the spectrum more suited to a cat's eyes than a human's, I see their shapes in the shadows. A breeze disturbs a stand of sedges and I imagine I see the rippling striped flank of a cat. A rounded, feline-shaped boulder seems to stir in the gloom. The place is full of these ghosts but my binoculars gather more light than my eyes can, and none of it is real. The dome of sky never quite darkens but by 11.30 p.m. it's hard to make out much at all on the contours of the landscape before us. I wonder how far away the nearest Scottish wildcat is right now. What's it doing? I imagine a mother wildcat leading a pair of kittens alongside a stream, searching for frogs or mice. Or a tom cat tucked into a hollow, pupils like saucers drawing in every grain of light the starless sky has to offer, still and breathless and watching patiently as a rabbit lollops closer.

The week's pattern is set. In the early morning I go walking alone in the hills and along the shores close to Kilchoan. In the middle of the day we go away to explore somewhere new on the peninsula. On our middle day, we head to the tiny port east of the village and take the ferry over to Mull. I stand on the deck in drizzle, watching Manx shearwaters scything past on their stiff, mini-albatross wings. After less than an hour's chugging, we draw into the dock at Tobermory. This cheerful town's colours are undimmed by the miserable weather. Mull is a wildlife mecca but not for the would-be wildcat-watcher – there are no wildcats here.

Or perhaps there are. In 2010, a tourist photographed what appeared to be a wildcat in northern Mull, and there are a handful of other anecdotal reports. The Scottish Wildcat Association (SWA) takes a circumspect view, suggesting that wildcats could theoretically manage the 1.6km sea crossing from the mainland – it's a bit of a stretch but it's also well known that wildcats are much stronger and more willing swimmers than domestic cats are. The SWA also puts forward the idea that wildcat-like hybrids that found their way to cat-rescue charities could have been brought to the island by people and ended up roaming and breeding.

Our short rainy walk on Mull, south-east from Tobermory, doesn't yield anything like a wildcat sighting, but we console ourselves with fish and chips. For people staying in Kilchoan, Tobermory is the nearest place to visit a supermarket, a cashpoint machine, and to hang out with lots of other humans. Which is all very well as far as it goes, but I am glad to board the ferry and return to the peace of Ardnamurchan.

The next morning, I see a cat. She's a skinny black-and-white almost-kitten who rushes up to me for a stroke as I'm starting out down the road into the village. I oblige and tickle her tummy, hoping that she has been neutered. Then she carries on westwards towards the rugged headland as I set off east to the village shop. Bumping into a friendly domestic cat is normally a high point of any day, but I'd rather not see them here.

Later on, we drive eastwards, down to the shore of Loch Sunart. This is the sinuous sea loch that separates Ardnamurchan from Morvern, and we've booked ourselves a boat trip to check out its wildlife. The boatman gives us our life jackets and helps us aboard a tiny boat that would be swamped by anything resembling a wave, but the loch is mirror-calm. Besides him and the three of us, there's just room for one more passenger – Tag the black Labrador, the boatman's dog, who sits quietly beside me and rests his big, heavy muzzle on my shoulder. We pull out into the deeper water and glide westwards.

Soon we are surrounded by moon jellyfish. I am fascinated by their graceful, mindless motion and their beauty, their translucent bell bodies revealing some internal arrangement of four connected horseshoe shapes that seem to glow inside (I learn later that these are their gonads). There are also occasional bigger jellyfish – *Cyanea capillata* – these ones opaque and red-tinted with a big, papery-looking bell and a trailing thick mane of hair-like tentacles.

The boatman guides us close to the shore to point out otter spraint (droppings) on a little jetty (Tag jumps off the boat here, goes over to the spraint for a sniff, and returns), and then the crunched-up remains of a sea urchin that was probably that same otter's breakfast. The conversation turns to wildcats, and the boatman tells us of a plan in progress to establish a wildcat population right here on Càrna, the little island that sits in the loch in between the two peninsulas. I can see the advantages. It would be easy to keep this island free of feral cats – it could be a safe haven where a few wildcats could live a natural life in very large enclosures. We head for Càrna and slowly circumnavigate it – it's rugged, steep-shored, forested. The boatman tells us that the 600-acre island is privately owned and has three houses on it – all of them are holiday lets, so there are no permanent residents. The proposed project, I discover, is part of something I have come across before: a wider and extremely challenging scheme to establish larger areas free of feral cats within this region of Scotland. This would give any remaining pure (or almost pure) wildcats here a good chance of expanding their population. The project was set up by the Scottish Wildcat Association in 2008 and now continues under the name Wildcat Haven. Something to research in detail when I'm home.

At one end of the island there are steep cliffs, and on them a few pairs of shags. These more maritime cousins of the cormorant are strikingly beautiful in their oily-glossed green-black plumage. They are serpentine, sleek and carry the remains of what would have been, a couple of months ago, a splendid quiff at the upper base of their bills. Now, with breeding well underway, these fancy feathers of courtship are redundant and have mostly gone. Though connected to the open Atlantic, the loch doesn't feel like 'real sea' as we're corralled by land on both sides, so it seems odd to see these true seabirds here. It seems even odder to spot a harbour porpoise, briefly surfacing to puff out a little plume of water before vanishing again into the depths of the loch. Terns – Arctic and common, are here too, skinny white-winged beauties that skim by in buoyant flight, dipping down to pick something from the water or sometimes just folding in their wings and dropping in face-first, surfacing with a wriggling fish clamped in their scarlet bills.

When the boat tour ends, we meander slowly back across the peninsula. This evening, and the evenings that follow, we spend as far from villages and houses as we can get, out in the wilds among the deer and the martins and the eagles, but tonight and every other night our evenings remain cat-less. The campsite keeps a book of wildlife sightings, and I check it each day, but no one's claiming a wildcat round here – no one has for months. However, on the day before we are to go I see that someone watched an otter in the sea that morning, right by the campsite, so I decide to make a very early start the following morning.

It's light well before 4 a.m., when I pick my way through the dew-sparkled tents and vans dotted across the main field. A little gate lets me down to the shoreline and then it's a difficult and slightly scary path down to the tiny bay where the beach widens and there's room to sit and wait. I stumble along between big rocks, trying to keep my big camera lens from bumping against them. Out on the beach, I find a flat rock and sit, look out into the bay, and straight away spot the dark whip of an otter's tail as it flicks upwards.

I've never had the chance to watch an otter at length before, but I count myself lucky to live in a time when I can watch otters at all because they were in dire straits just a few decades ago, their numbers laid low by devastating river pollution from the 1950s. Prior to this they were hunted and much persecuted as well, but use of organochlorine pesticides brought about a further, dramatic decline which continued into the 1970s. Since the 1980s, otter numbers have been recovering, to the point that they are now easy to see in several parts of England, but north-west Scotland remains by far the best place in the UK to find them. Their recovery mirrors that of the other predatory mammals that were very rare in my childhood – the pine marten and polecat are also on the upswing. The sole exception to this trend is the Scottish wildcat.

The otter is cavorting, swimming on the surface with much tail-flicking, arching and rolling. Every so often it dives and is underwater for a couple of minutes – I take this time to move down the beach a little and find a new vantage point slightly nearer the shore. In this way I get closer and closer to the water's edge, while remaining (hopefully) inconspicuous through my immobility while the otter is at the surface. The light is not good and most of the otter is hidden underwater the majority of the time, but I take pictures anyway – of a tipped-up tail, a humped-up spine, and a blunt, silhouetted head haloed with sunrise gold.

Then the otter surfaces with something in its mouth and strikes out towards the shore. My chest grows tight as I realise it is heading directly towards me. What it's carrying is a big crab with a mass of spiky legs that look like an extravagant moustache sprouting from the otter's face. It reaches some low rocks right on the shore, about 4m away, and hauls out its sleek, slick body. It positions itself side-on to me and sets about demolishing the crab, sending sparks everywhere – water droplets and shards of carapace flying around its face. At one point it pauses

to shake itself and when the starburst of spray settles it's
no longer sleek but spiky, its curves and edges delineated
in a geometric pattern of soft spines. I see all this through
my viewfinder as I take photos as fast as the camera
buffer can manage. At some point my memory card fills
up – I change it as quickly as I can and carry on.

Having devoured all the best bits of the crab, the otter
leaves the remains behind and eases back into the water.
When it submerges I leave my rock and walk across the
beach to find another vantage point. I'm just settling
here when I realise, with a sick jolt, that I don't have the
memory card I just took out of my camera. I think back
to what happened. I'd put it on my leg while I was
swapping in the new card, with a view to pocketing it,
but then I moved from my spot and forgot it was there. I
just stood up and it must have fallen off without me
noticing. I retrace my steps – or I think I do – but it's not
where I thought I had been. I search as patiently as I can,
wondering how I could lose myself so easily on this
relatively small segment of beach. It must have been only
five minutes of hunting before I spot it, a bright orange
square glowing cheerfully among the slate-grey rocks
and gravel. Giving thanks that it hadn't landed in a
puddle, I grab it and look back out to sea. No sign of the
otter. It probably saw me moving around and decided to
beat a retreat. I should have kept an eye on it and kept
still when it surfaced. Fear of losing my images made
me careless.

Feeling foolish and slightly disappointed, but also
dizzy with relief, I sit down again, let my breathing settle,
and take a slow look around the beach in the brightening
sunlight. A movement nearby catches my eye. There on a

ridge of rock close by sits a dusky-grey bird – a fledgling rock pipit in fluffy but freshly perfect juvenile feathers. It is hopping about on the top of the tallest rock in a desultory manner, but then drops into a crouch and begins to call. A second 'rocket' arrives – an adult. It looks in a slightly sorry state compared to its baby – its plumage sun-bleached and disordered. It carries a mouthful of indeterminate squashed insects, and it gives me a cautious glance but decides I am not close enough to pose any danger. It hops up and posts the unappealing-looking meal into the chick's gaping pink mouth, and then flits smartly away to fetch more.

I watch the pipits a little longer. I look out to sea where distant gannets, auks and shearwaters flap purposefully by, and closer at hand where oystercatchers and ringed plovers trot across the shoreline rocks, and where a grey seal lolls in the sea, its muzzle a snorkel pointed skywards as it drifts and sleeps. Three hours have gone in an instant. It's the story of the week – those long, long hours of daylight that race past so quickly, slipping through my hands like Sanna sand, like an invisible wildcat padding away into a black forest. Time to go home.

CHAPTER FOUR
Life Unseen

Scottish wildcats famously shun human contact. It's sometimes said that if you see a wildcat, that's proof in itself that it isn't actually a wildcat but a convincing hybrid whose pet-cat genes are the only reason it let you lay your eyes upon it in the first place. It stands to reason that an animal with such a long history of persecution by humans should give us the widest berth it can – a hundred years ago, any Scottish wildcat that was tolerant of human presence would probably not live long enough to pass on its approachable genes to the next generation. In contrast, while feral cats are often quite fearful of people, they are nowhere near as worried by humans as Scottish wildcats are, and may choose to live in close proximity to us. Scottish wildcats, though, are most likely to be found in very remote, wild areas where the chances of bumping into humans are minimal.

Because feral cats are less human-avoidant, they sometimes live a not-quite-wild life and will accept handouts from people. If this food source is abundant they can form large colonies, living in much higher densities than is normal for any kind of wildcat, and within such colonies a sort of social structure will develop. Scottish wildcats never show that sort of sociality. This is perhaps the most important difference between the two, but when Scottish wildcats and feral domestic cats interbreed, the resultant hybrid young are likely to show a mixture of behavioural as well as physical traits, blurring the lines between wildcat and feral cat behaviour.

Although they are so similar in so many ways – similar enough to breed together, for a start – the fundamental difference in how they feel about humans means that the lives of 'pure' Scottish wildcats are much less easily observed than those of feral cats. What we know about how Scottish wildcats live in their natural environment is minimal because of their rarity, the sheer difficulty of finding them (let alone watching them at any length), and the fact that most of them today are likely to carry at least some domestic cat genes. However, we can infer certain things about them by other means: historical notes, observations of wildcats in captivity, observations of free-living European wildcats in other countries, and (more tenuously) the behaviours of feral and domestic pet cats.

While the differences between wildcats and tame cats are important, they share plenty of commonalities too.

All Scottish, European and African wildcats, and all pet and domestic cats, have to eat meat to live, and their most basic instincts tell them that the way to do this is through hunting. They also need to avoid falling victim to other predators that are bigger and stronger than themselves – by hiding or fleeing as the initial lines of defence, and by fighting as a last resort. For male cats in particular, being able to fight and defeat other cats is also an important skill. Even the most pampered pedigree cat, descended from a long line of pets as indulged as itself, will retain at least some of this primal nature, and only those wildcats with the best hunting and fighting instincts can survive.

It is the ability to take down a wide variety of vertebrate quarry – many of them being animals with considerable power and intelligence of their own – and the stalk-and-pounce way that cats put their skills to work in the hunt that shape the basic feline personality. The saying goes that curiosity killed the cat, but I'd prefer to put it like this: cats are courageous. They are naturally inclined to approach, explore and investigate, and to rush in headlong when they commit to an attack. There's individual variation, of course, but generally wildcats and domestic cats alike are extremely brave, especially given their small size and the fact that they are legitimately prey for a handful of top predator animals as well as being predators themselves. In dire straits even the most mellow pet cat will fight back with utter ferocity.

Tame pet cats with no anxieties around people show us the hunter side of their characters in the most charming way – through play. Before they can even walk or run properly, kittens start to attempt to pounce on and seize any object of interest – including their own siblings. By the time they are a few weeks old and mobile, play dominates their waking hours. Attacking objects of all kinds with claws and teeth is predatory behaviour, and rough-housing and stalking its littermates are behaviours that help to teach a kitten how to tackle a victim that might fight back. It is a vital part of kitten development, and the timespan between their being able to play with full physical commitment and their becoming independent of their mother is short. Within that timeframe, every wildcat kitten has to learn how to hunt well enough that it will be able to kill all of its own food without any help, especially through the winter when quarry is scarcest. For most pet kittens, there will never be a need to actually do this, but the instincts to act in this way remain strong and never fully disappear. Even a doddery old adult cat in its late teens can often be drawn into a game with a bit of string or a scrunched-up ball of paper from time to time.

Play with a cat and you'll see the way it hunts. As soon as the object of interest is jiggled around, the cat goes into stalk mode, crouching and freezing with eyes fixed on the target and pupils dilating. The crouched position will power the attacking pounce, driven by the hind legs as the cat springs forwards, front paws outstretched with claws unsheathed. The cat can cover about a metre in its final pounce.

If you swish the string or roll the paper ball close enough to the cat that it doesn't need to leap, it will strike very quickly with one paw from where it is, attempting to pin the target on the ground. The strike speed of a Scottish wildcat making a grab in this way (based on footage of a captive wildcat) is about 1/60th of a second – much quicker than that shown by a domestic cat.

Flick the string in the air, or throw the paper ball high over the cat's head, and you may well tempt it to try a vertical leap. It will make a swipe or grab in mid-air, the way that it would attempt to bring down a bird in flight, and its accuracy is impressive. Whichever way the cat tries to strike or grab with the paws, if it successfully takes down the target, it will quickly move in to bite what it has caught. It may then roll over with the 'prey' held in its front paws, and start to kick at it forcefully with the hind feet, a behaviour that is seen during fighting as well as when trying to immobilise big prey.

This is the way Scottish wildcats hunt and catch the animals they eat. They sit and wait, perhaps for hours in some situations (when staking out a rabbit warren, for instance). Alternatively, they move in a slow walk, but in both cases they are constantly watching and listening for movement of quarry. If they detect a potential victim, they drop into a tense, hyper-alert crouch, and slink closer if they feel they can, each step very careful and deliberate, their eyes locked on the target. If they can, they move along landscape boundaries for extra concealment, such as vegetation, drystone walls, or just natural ridges and dips in the landscape – their pattern of broken dark stripes provides good camouflage against many natural backdrops. When close enough to strike, they crouch low and may perform the familiar cat bottom-wiggle to line up the hind paws on firm ground, ready to pounce. However, if a target appears suddenly, they can also react at lightning speed with a spontaneous dash or a leap.

Once a cat has hold of a victim it will seek to deliver a quick, hard bite to the neck, which will be enough to kill or at least disable all but the biggest and toughest quarry. A Scottish wildcat is most likely to perform the roll-and-kick tactic when struggling with a large rabbit, mountain hare or other sizeable prey item – the vigorous kicks and raking strikes of the hind claws can quickly disembowel the victim.

Both wild and domestic cats are versatile hunters. They do best at hunting small and smallish mammals, from mice and voles up to rabbits, but they are also able to catch birds within a similar size range and will readily have a go at reptiles and amphibians too. They can overcome their aversion to water if there's a chance of scooping out a fish or newt from a pond, and they will also capture insects and other invertebrates provided these are moving fast enough to trigger the stalk/pounce response. Cats will, of course, also eat food that's not moving at all if it smells appealing. Pet cats don't need their food bowls to move about enticingly – the aroma is enough. And if a hungry Scottish wildcat happens upon a bird's nest it will recognise that it's found food by the scent and will eat the contents whether they be eggs or chicks. Ground-dwelling birds that rely on camouflage are also vulnerable to being unmasked by the keen wildcat nose. And, although wildcats are highly predatory, they will also eat fresh carrion.

Some pet cats are extremely adept hunters of all kinds of quarry. Some become specialists, attacking birds in preference to rodents or vice versa. Individual Scottish wildcats tend to be specialists too, taking whatever is abundant in their particular region or habitat type. There is a marked east–west split in the general make-up of the wildcat diet, with those in the east taking far more rabbits than those in the west, which eat mainly small rodents.

The 'best' prey for a wildcat is a rabbit or a hare (brown or mountain), as these large animals deliver plenty of meat for a single hunting attempt. Of those three kinds of lagomorphs, only one (the mountain hare) is native to Britain, and we know it now as a rather rare animal of the highest uplands, its distribution not overlapping that much with wildcat habitat as things stand today. However, it was probably much more widespread historically – before the rabbit and brown hare were introduced to Britain and outcompeted it in the lowlands – and would have been a key target. Tackling the bigger, faster brown hare is a challenge (albeit not an insurmountable one) for a Scottish wildcat, but rabbits are relatively easy prey and their habit of living and breeding in large, conspicuous warrens also helps make life easier for predators like wildcats.

One big wildcat needs to eat the flesh of a couple of rabbits a day to survive, but if it is subsisting only on smaller mammals it needs to hunt much more often – some 25 to 30 mice or voles a day will be needed to meet its energy needs. The small rodent most prevalent in wildcat country is the short-tailed vole, but there are also bank voles and wood mice in abundance in forests and forest edges, and common and pygmy shrews in all kinds of habitats. Less frequent mammalian prey that has been recorded as taken by wildcats includes moles, red squirrels, brown rats, and roe deer fawns. Whether wildcats can kill small lambs is a contentious point, but the possibility is enough to have earned them an unfavourable reputation among shepherds.

Cats are more suited to catching mammals than taking birds, but they can become skilled bird-hunters; indeed, the Scottish wildcat will certainly not pass up any opportunity to try to catch one. Small passerines like pipits and chats are taken frequently, as are larger quarry such as woodpigeons, ducks and pheasants. Species that nest or feed on the ground, such as golden plovers, woodcocks and game birds, are more vulnerable, and although wildcats tend to avoid the open heathery landscape of grouse moors, they will prey on red grouse if they happen upon them, and so historically their presence on grouse-shooting estates is not readily tolerated by gamekeepers. They may also overcome their aversion to human presence in order to raid a badly secured hen-house.

In some areas, small reptiles like slow-worms and common lizards are frequently taken; likewise frogs and newts. Any insect prey is likely to be the bigger, more conspicuous species, such as large moths, dragonflies or ground beetles. The Scottish wildcat's willingness to take carrion is exploited by researchers using camera traps to search for their presence. A meat bait, such as a rabbit leg, is placed up on a post, so that the wildcat has to stand on its hind feet to reach the food. This means the camera facing the post should get a clear view of the tail pattern, which gives at least some idea of whether the cat in question is feral or a candidate for a pure Scottish wildcat.

What happens to a wildcat that can't hunt, or can't hunt well enough? Starvation, most likely, but there could be another option. When I was growing up, we often visited a local friend who lived in sheltered accommodation in a semi-rural area, and the surrounding woods were home to a colony of feral cats. Our friend fed the cats, as did some others in the accommodation. We ended up adopting a kitten from the colony, which our friend had identified as being particularly inept at fending for herself – this feral-born kitten grew up to be an exceptionally docile cat. Many of the others remained wild and fearful but they all still came to take food close to the residents' homes. From what I observed, actually hunting for their own food was rare in this particular colony. I've heard stories of Scottish wildcats visiting rural gardens or loitering around country hotels and taking food put out for them, but whether a pure or nearly pure wildcat would ever scavenge in this way seems improbable. However, a hybrid wildcat–domestic cat might well overcome its wildcat side and linger around houses for scraps if the alternative is starvation.

As things stand, Scottish wildcats are top predators in Scotland. A young wildcat kitten might be killed by a fox or a buzzard, and there have been occasional reports of wildcats doing battle with golden eagles and coming off worse. However, historically there were Eurasian lynxes, wolves and brown bears living in wilder parts of Britain, and all of these would potentially prey on wildcats. In fact, there is a good chance they would particularly target wildcats, thanks to the curious phenomenon of intraguild predation. This behaviour describes the tendency of predators to seek out and kill other, smaller predators that represent competition for resources. The Eurasian lynx's diet and habitat needs are not dissimilar to that of wildcats, and observations from mainland Europe where both species occur do indicate that the presence of lynxes is associated with lower-than-expected numbers of wildcats. A lynx is certainly big, fierce and strong enough to kill a wildcat, though the fight would be ferocious indeed. Intraguild predation is an important phenomenon in ecosystems – in Britain, for example, it accounts for the sparse distribution of the long-eared owl, which is victimised by the larger tawny owl where both species occur together. With reintroduction schemes in the works for both Eurasian lynxes and wildcats, the impact of the one species upon the other must be taken into account.

If you have two kittens or cats that play together, you'll get a look into how cats in the wild might deal with attack by another predator or a territorial rival. As the attacker rushes in, the victim will try to race away, using all its powers of acceleration. If the attacker corners the victim, a stand-off may ensue, with the two face-on or side-on to one another, ears back and backs bristling, displaying their bodies in as intimidating a way as possible. Each is sizing up the other and looking for an opportunity to spring (the attacker is likely to do this first but the victim will often attempt a pre-emptive strike). When one jumps on the other, they grab on and roll around together, trying to bite at each other's heads and necks and kicking at each other's bellies. Pet cats that live together and have a cordial relationship engage in silent, restrained play-fighting, with only the occasional yelp of alarm or pain if things go too far. But when true rival domestic cats or wildcats fight each other, the bites and the kicks are full-blooded, intended to cause real damage. Their stand-offs are accompanied with low growls and hisses, and their physical struggles with loud and nerve-jangling shrieks.

We usually remove pet kittens from their mothers at eight weeks old, sometimes younger, but wildcat families remain together for up to six months after the kittens are born. The extra time they have under their mother's care is vital – not only for honing their survival skills but for them to grow to a more survivable size where they are less likely to be attacked by other wildcats or predators. But eventually the mother wildcat becomes hostile to any kittens that linger too long, as it is time for her to become pregnant with her next litter, so the kittens she already has will have to leave her territory and seek out their own. By this time they should be able to hunt well, but they are still small, vulnerable, inexperienced and – worst of all – without a territory of their own. Getting through its first winter is a huge challenge for any young wildcat.

Scottish wildcat habitat is, as mentioned already, usually very remote, in places where encounters with humans are highly unlikely. It also needs to have a population of prey, and some safe, sheltered place (ideally more than one) that will serve as a den in which the cat can sleep, take cover from bad weather and (if female) have kittens. Habitats that are most likely to support them will most probably be in the uplands and will combine areas of pine forest or birch woodland with more open grassy moorland or other open country – the border between forest and grassland is often a particularly productive hunting zone. Boulder piles or spaces around the roots of large or fallen trees can offer sites for dens, as can the burrows dug by other mammals – old fox earths or rabbit burrows may be used. Complex, rugged landscapes with streams and patches of bushes offer more potential for stalking and hiding, as well as less chance of encountering roaming humans. Wildcats may find denning sites in more densely vegetated habitats – a bit further into the forest, for instance – but will need more open areas to see prey and make a successful hunt. They mostly avoid dense heather moorland, the deep interior of coniferous woodland, and open agricultural land.

No habitat, however promising-looking, will support wildcats if it doesn't also support prey. However, historically many Scottish wildcats have long lived in a landscape in which quarry is fairly sparse and so they are adapted to use a large area for hunting. Some in the west Highlands may range over an area as large as $25km^2$, and cover more than 15km on foot in a single night in their search for prey. However, in the east of their distribution, where quarry is more abundant, the home range may be smaller than $2km^2$. A stable central or core part of the home range is defended vigorously against intruders – this might be defined as the territory proper – whereas the outer reaches of wildcats' home ranges may change in extent frequently and may overlap with the ranges of other wildcats. A male's home range is especially likely to overlap with that of one or more females, although social interactions are non-existent until it is time to mate.

Both wild and domestic cats mark their territory with poo and pee in order to leave a long-lasting message to other feline passers-by. When a cat excretes faeces, this squeezes out the contents of the anal glands, which add their distinctive aroma to create the bouquet of unpleasantness that cat owners and many gardeners know so well. Cats of both sexes also spread scent with urine, though males can do so more fulsomely as they stand and spray rather than squatting as females do. The final piece of the fragrant puzzle comes from scent glands in the cheeks and paws: the cat spreads these smells by rubbing its face on objects and by scratching at them, respectively. Cats patrol their territories regularly and scent-mark the same prominent points repeatedly, especially when other cats are around. Domestic cats – particularly those that are subordinate within a dense cat population – do bury some of their poo to try to hide their presence. However, poo-burying is not the norm for Scottish wildcats, except perhaps very close to the den area.

With Scottish wildcats now so rare and sparsely distributed, the problem of finding a suitable area to claim as a territory is perhaps less serious than it would have been a century ago. Yet territory still remains tremendously important. Wildcats are homebodies, not nomads, and to survive they rely on an intimate knowledge of their home patch. Experience will teach them how to make the best use of the space, exactly when and where the best hunting opportunities will be, and the key points to monitor for possible mating opportunities or intrusions from neighbouring territory-holders.

Unneutered female domestic cats (or 'queens') come into heat about every two to three weeks through spring and summer, beginning at about five months old and continuing throughout the adult life. Most of them have a breeding season dictated by day length (only coming into heat when days are longer than nights), so this shows some regional variation. However, some individuals may come into heat all year long regardless of where they live. If she is allowed to mate at will, a queen will have about three litters a year. Scottish wildcats, by contrast, naturally breed just once a year because the kittens remain with their mothers for about five or six months. A wildcat queen would, in theory, have time to have a second litter if her first is very early and she then loses her kittens while they are still young, but there is no evidence of this happening in the wild. Feral cats living in colonies may show shared maternal care, something that is not documented in Scottish wildcats (indeed, females are fiercely territorial against one another). This behaviour – whether natural or artificially induced by an abundance of non-natural food – can enable some of the queens to have two or three litters in a season.

A pet cat on heat is hard to ignore. She will call constantly, and this vocalisation is nothing like the familiar meow used to get human attention – it is a drawn-out, throaty yowl, an altogether more primal sound. She will roll about on the floor restlessly and walk in a curious waddling, half-crouched stance – if you stroke her, she will probably move her tail to the side and lift up her rear end. The general impression is of an unhappy and agitated cat. She is also releasing pheromones to attract unneutered males ('toms') through strong-smelling urine. She will stay in this condition for a few days, and then will return to normal until the next heat cycle begins, unless she mates. If she is allowed access to the outdoors while on heat (and she will do all she can to get outside) it is almost certain that she *will* mate – there are free-roaming or feral tom cats around almost everywhere in Britain and they will travel miles in search of queens on heat.

The call of a female wildcat on heat is historically described as a horrific screech, and males trying to get close to her while warning each other away is similarly unearthly. It is easy to imagine how alarming this cacophony would have been to shepherds and crofters out on late winter and spring nights. The cries of both sexes in the throes of this courtship ritual will attract additional males to the area, and they will also try to get close enough to the queen to attempt to mate.

As an aside, the sounds that we most associate with our pet cats are not typically made by wildcats. The plaintive mews and meows that cats give are for our benefit – a kittenish habit that never goes away. Watch cats interacting with one another and you are unlikely to hear a single mew, but when a cat wants your attention it will say so in this way. Wildcats will purr, and mothers will chirrup to their mewing kittens, but you probably won't ever hear an adult wildcat meowing – even feral cats rarely do.

Courtship behaviour begins in late winter, with the female on heat for about five days at a time. The females will typically be pregnant by March at the latest, and gestation lasts 65–69 days – up to a week longer than in the domestic cat. Mating is a cagey and sometimes violent affair in both domestic cats and wildcats. Eager though she is to mate, the female is wary of allowing the larger, stronger male that close to her and he will often be repelled with snarls and swipes on his initial approaches. When she finally accepts him, she allows him to rush at her and seize the skin of her nape in his teeth. Then, once he has got a grip on her neck, he mounts her. The act of withdrawal is painful for the female as the male's penis bears forward-pointing barbs, which rake her inside as he disengages. Her reaction is to shriek loudly, struggle out of his clutches and often then to attack and drive him away. Cats do not ovulate until they have mated – ova are released from the ovaries 20–50 hours after mating. The female may mate with two or more different toms while on heat, meaning that her litter could have more than one father.

Litter size is one of the ways in which domestic cats and Scottish wildcats differ. It is unusual for a wildcat queen to produce more than four kittens, and litters of two or three are frequent. Domestic and feral cats, on the other hand, have four or five kittens per litter on average, and more mature queens will have six or seven quite commonly. Bigger litters are common in certain breeds, such as the so-called oriental types (Siamese, Burmese and related breeds). The largest domestic cat litter on record is of 19 kittens (15 of which were born alive) born to a Siamese–Burmese cross queen. Having larger litters and regularly breeding twice or more in a season means that feral cats can easily outpace Scottish wildcats when it comes to building a population.

The pregnant female wildcat needs a den in which to bear and nurse her kittens. She will probably have more than one option in her territory. She does her best to keep the den's location hidden, leaving minimal traces of use in its vicinity. The inside of the den is bare.

Birth in free-living Scottish wildcats is, understandably, not really documented at all, and is seldom observed in captive wildcats either, as it is best that keepers give the queen space and peace when her kittens are due. But we can have a fair idea of what happens from observations of domestic cats. The birth is quite a protracted process – signs of labour can be evident more than a day before the first kitten is born. The queen will be restless and will pace around inside her den, in between lying down and licking herself. She will pant and generally look uncomfortable. When the first kitten finally arrives, she will lick it clean of its amniotic sac, paying particular attention to its face. The grooming also encourages the kitten to begin to breathe. She will chew through the cord and often eats the placenta. The kitten will soon locate a nipple on the queen's belly (most queens have six or eight nipples) and begin to suckle. The rest of the litter will follow at variable intervals – usually of up to an hour between each birth but sometimes longer. Each kitten finds a nipple from which to suckle and will always try to latch on to the same one for all subsequent feeds.

Newborn kittens, both wildcat and domestic, are furry with their coat patterns already apparent, but the eyes are closed and they are weak and almost helpless, able only to crawl a short distance and to make a surprisingly loud squeaking call. They huddle against their mother and suckle intermittently; when she has to leave them to find food they remain piled up together until her return. Scottish wildcats time their births for the point in the year when natural food is at its most abundant, making the queen's foraging trips as short as she can manage – even so, she will need to eat much more than usual while she is suckling young. Her daily calorie requirements double, triple or even quadruple as the kittens grow, depending on the size of her litter.

The father or fathers of the kittens will be long gone by this point, so the mother wildcat has to handle all of this on her own. There are a few anecdotal accounts of male Scottish wildcats (wild and captive) associating with females that have kittens and even providing food for them, but this seems to be very much the exception. The mating system of wildcats is such that no male can be especially confident in his paternity, so it is not wise for him to invest time in supporting kittens that may well not be his. There are records of wildcat toms actively seeking out and killing kittens – presumably in cases where they have had no contact with the mother and so can be certain that they are not destroying their own offspring.

It will be ten to 13 days before the kittens' eyes open. The newly opened eyes are pale blue, but gradually change to the adult-like yellow–green over the next nine weeks. At about a month old the kittens will be taking their first mouthfuls of solid food and waddling around the den on rapidly strengthening legs. By six weeks or so they can run and leap, albeit clumsily, and at ten weeks they will start leaving the den with their mother when she goes out to hunt. By 12 weeks old they will be eating meat exclusively, and catching and killing some prey for themselves.

Hiding and guarding the young kittens is a huge and difficult undertaking for the mother wildcat. When they are tiny she keeps evidence of their presence to a minimum by eating their waste, and by the age of just one week the kittens are already sufficiently wildcattish that they will hiss and spit at any intruders at the den, hopefully putting on enough of a discouraging display to hold off danger until the mother returns. There are myriad accounts of wildcat mothers ferociously attacking dogs, foxes and humans that have threatened their young. If she successfully fends off an intruder at her den, the mother will probably move the kittens to an alternative den as soon as she is happy that the coast is clear. When she moves them, she carries them in her mouth by the loose skin on their neck, and the act of being seized in this way activates a reflex which makes them relaxed and compliant. In domestic cats, the 'go-limp' reaction on being 'scruffed' is retained into adulthood, but no one has been brave enough to see if the same is true of wildcats.

It takes a brave predator to confront a protective mother wildcat, but kittens are sometimes taken by opportunistic hunters, such as foxes or golden eagles. As the kittens grow older and more independent, they are more likely to wander an unsafe distance from their mother's side. Their natural curiosity can often lead them into trouble. About half of all wildcat kittens don't make it past four months old, despite the constant and committed efforts of their mothers to keep them fed and safe.

Young wildcats' first hunting forays, alongside their mothers, involve stalking and catching moths, beetles and other insects – relatively hapless and easy prey. However, these expeditions also offer opportunities to watch their mother taking on more difficult quarry, and when she captures a rodent or other larger victim she may not kill it but offer it to the kittens as a living toy for hunting practice. The kittens can pounce on a wounded mouse many times as it tries to escape, developing vital skills.

By the time they are five or six months old the kittens will be starting to move on. They may do this of their own accord or be encouraged on their way by their increasingly unfriendly mother. This is the hardest time of their lives, with only fledgling hunting skills, winter on the way and no territory of their own. They are still small enough that they might be bested by another predator, and on their wanderings they'll also face new hazards, such as road crossings and perhaps other cats whose territories they may accidentally enter. Sometimes a mother cat will tolerate one of her kittens staying in her home range for longer than the usual period – especially female kittens. The males, which require a bigger home range as they need to be in contact with as many females as possible, are more likely to leave earlier and to wander further.

It is probably these young wandering males that are most likely to meet wandering female feral or domestic cats in the late winter and to father litters of hybrid kittens – they can breed at just nine months old. We can't blame a young wildcat for taking whatever mating opportunities he finds, particularly when he is so much more likely to meet cats of domestic origin rather than other wildcats. However, here in the twenty-first century, interaction with domestic cats threatens the Scottish wildcat's existence more than anything else.

TRIP THREE
Speyside, 2014

I make another trip to the Speyside area in November 2014, once again boarding the overnight coach to Aviemore and settling in for 12 uncomfortable hours of riding through the dark. The days are short now, and we are well into Scotland by the time the sun rises. My excitement at returning to my favourite Highland valley is tempered by the knowledge that I'm searching for the finest of needles in the hugest of haystacks.

Speyside, also known as Strathspey, is one of the 'priority areas' identified as holding (comparatively) high numbers of wildcats, although it is also an area where hybrids are likely to be more frequent, given the relatively high human population here. It's more accessible than most of the other priority areas, especially if you get about by coach, train, bus and feet.

Timing a trip to look for Scottish wildcats is difficult in a way, but in another way it's not: no matter what time you go, you're almost guaranteed not to see one. But in theory, late autumn is a time when the population of individual roaming wildcats is at its highest, as those kittens that made it to independence will have moved out of their mother's territories and will be seeking a patch of their own. They will be covering many miles each night and may be exploring new areas, possibly closer to human habitation than usual, in their quest for a productive territory. Summer's lush vegetation will have died back enough to make them easier to see as they do so, too. My relentless optimism – a mixed blessing for a wildlife-watcher – surges to the fore and I pass the twilight hours on the coach in a mood of sleepy good humour.

A pause in Aviemore, a bus ride and a half-hour walk and I'm back at Dell Cottages in Nethy Bridge again – in a different one this time but, like last year's, it has a little square of garden with trees at the back and a bird-feeder that is already busy with chaffinches and coal tits. I dump my bags, grab my camera and binoculars, and head past the last few fields between here and the forest. The last field of all is a spot where I always pause, as it stretches away into the distance, meeting woodland edge, with the grey bulk of the Cairngorms brooding beyond. The field itself is lusciously green, wet and sedgy, the sort of place where rabbits and rodents would thrive and wildcats could hunt them. So I stop, and scan back and forth. Then, overhead, a noisy storm of jackdaws erupts and they take my attention upwards. The flock spreads and fans out over me and I notice how, even within their massed morass, they are flying in pairs – each independently moving element of the flock is a twosome rather than an individual bird. I remember that I have read that jackdaws are unusually faithful birds, shunning all opportunities for what biologists call 'extra-pair copulations'. Most other birds, although they appear monogamous, will mate with a neighbour's mate if they can get away with it. Not jackdaws, though. Such is their devotion that even in the air they need to stay close to their partner.

It gets me musing about sociality in animals, in people, in my own nature. I'm up here alone for a fortnight – if I talk to any other people it'll be only in passing. Most days I'll probably speak to no one. As an introvert with more than my share of autistic traits, I do solitude pretty well, but I'm still a human and I do feel that pull towards my own kind sooner or later. So I sit somewhere between a jackdaw, who always has to be with its flock but, even more than that, has to be with its lifelong partner, and a wildcat, who has to be alone forever, for all but the most essential biological reasons.

When I was at school, one of my classmates had a pet jackdaw that waited for him outside the classrooms while he was in his lessons. It sat restlessly on a railing at the door and, as soon as the boy emerged, the jackdaw jumped up onto his shoulder to press itself against his neck in palpable relief at the reunion. I was impressed and a little envious of their bond, but I wondered whether the jackdaw suffered agonies of loneliness when they were apart. Walking into the wood by myself on this quiet grey afternoon, I feel more kinship with a wildcat, but I know that a wildcat will never ever feel any kinship with me.

After ten minutes of slow strolling down the broad, main pathway through the pines, I turn left through a gate and into more open, rough ground. Here the path becomes indistinct, but I follow a climbing ridge alongside little stands of silver birch. The ridge curves and I'm overlooking a big, boggy field in which roe deer are grazing. Roe deer rut earlier in the year than our other common deer species, in late summer, and hostilities are long over now. There are three in the group before me, all does, and all have probably mated. They are not technically pregnant yet, though – the fertilised eggs they carry will not actually implant until after midwinter, with twin or triplet kids born in early summer, five months later. This anatomical trick means that they can mate before winter sets in, and go through the most physically challenging stage of pregnancy after winter is over. The implantation delay means that, during the leanest months, they minimise extra demands on their bodies.

Our other deer don't do this, but several other kinds of mammals with relatively long gestational periods do. It's known in some mustelids, including badgers, pine martens and stoats (but not weasels, with their brief five-week pregnancies). It is also the norm for the bats that live in Britain, for example, with the males copulating with sleepy females just as hibernation is about to begin – despite their small size, bats have very long lives and very long pregnancies. Wildcats, though, don't show delayed implantation. They mate in mid- or late winter, going through the physically demanding process of courtship (including fighting rivals) at a time when the stresses of life are already acute. By this time, the tom kittens of the previous summer are old enough to at least make an attempt to mate, and because they are also dispersing and possibly are well away from habitats occupied by other Scottish wildcats, their chances of meeting a feral or domestic female cat on heat are considerable.

I stop up here, finding a spot of dry ground to sit on, and watch the deer and scan the margins of the field repeatedly. It's mid-afternoon and the sky is already a touch dusky. This is one of the downsides of being here so late in the year. In my sleepy state I forgot to bring my

head-torch, so I have to head back before real darkness
takes over.

The garden looks so beautiful in the morning. It is
still overcast but this seems to heighten all the colour –
the lawn, the lichen-crusted birch twigs star-studded
with overnight raindrops, and the intensely copper-red
dead leaves of the beech hedge. They match the pelage
of the red squirrel that bounds across my lawn, then up
into the beech tree where the peanut-filled bird-feeder
hangs. The little birds scatter at its approach. It anchors
itself to the twig from which the feeder hangs, gripping
with both hind feet, then lowers its upper body like a
gymnast until it's hanging at full stretch against the
feeder. It grips the wire mesh with its forepaws and
works away with its teeth to pull a peanut through a gap
in the wire. Having succeeded, it holds the nut in both
forepaws and nibbles as it dangles, showing off a neatly
outlined white tummy. It is heading towards its winter
coat – its ear tufts are growing longer, its flanks and tail
becoming greyer – but it is still essentially fiery red,
autumnal.

When the squirrel finally climbs down and bounds
away, a great spotted woodpecker turns up, a flurry of
black, white and red, screeching to a halt somehow as it
lands on the birch trunk. It's a female, her nape uniformly
black (a male would have a little red patch there). She
scrambles up to the branch where the feeder hangs, and
then down, bottom-first, onto the mesh, deploying her
sturdy bill to bash away at the nuts. For every fragment
she eats, another dozen spin to the ground, where they
will be hoovered up by ground-feeders such as dunnocks
and robins. I'm going to need more peanuts.

The garden show is compelling but I need to be out beyond all of this, so I head off for a full day's walk, or prowl, around the forest and its edges. This forest is huge and feels particularly wild and beautiful on this blustery grey day. The Scots pines grow as they wish, giving the lie to our idea that, of our native trees, only the deciduous ones become enormous, spreading and bountiful. The trunks and lower branches are so thickly crusted with leaf-like lichens, and draped so heavily in trailing, cobweb-like lichens too, that it's hard to see any of their bark at all in some places. The lichens are complex organisms, a symbiosis of fungus and alga that are inextricably intertwined. And among the complex physical structure of miniature ridges and hollows, strands and plateaux that is formed by the lichen and the bark on which it grows, a whole ecosystem exists, a community of miniscule invertebrate life forms. These Lilliputian creatures sustain the little birds – the crested and coal tits, but particularly the goldcrests and treecreepers. Tiny birds with tiny fine bills, these two are adapted to probe and explore this miniature landscape to make their living. Most of the other small, exclusively insect-eating birds that live in Britain are summer visitors, heading hundreds of miles south in winter because they'll simply starve if they try to stay.

I watch a goldcrest at work. This is Britain's smallest bird species, weighing barely 6g, and it has to eat nearly non-stop through the day to sustain its tiny body – it is literally a matter of milligrams away from starvation at any given moment. Given that, this one seems to be remarkably profligate with its extremely strict energy budget. It's hovering like a funny-shaped hummingbird around the fluttering tresses of lichen hanging from the small birch by the pathside, snapping away at things so minute I have no chance of even seeing them. It pays me no mind as it lands in the tree and continues to investigate the lichen as it flits and bounces from twig to twig. Through the binoculars I glimpse its face, wearing a permanently sorrowful expression thanks to a huge dark eye and a dark cheek stripe that suggests a downturned mouth.

Finding a treecreeper is more difficult because they are less obvious, but I scan a few trees nearby and soon locate one, scrambling its way skywards in a broad spiral up the trunk, its marbled back camouflaged against the ashy and tawny colours of the crusty bark. It's hard to imagine a duller life for a wild animal – climb to the top of a tree, fly down to the bottom of another tree, repeat, all day long. But it works – treecreepers are very widespread in Britain, and their numbers have been pretty stable for some 40 years. The same goes for goldcrests. What long-term trends don't show is that these diminutive birds get absolutely hammered in severe winters – their numbers can drop by as much as 80 per cent if there's a lengthy freeze-up. Eating fast enough to keep their fierce little metabolic fires burning in sub-zero temperatures is often impossible. This is when these birds might, in desperation, start checking out garden bird-feeders. But those that do make it to spring after a terrible winter will enjoy more space, more food and consequently more breeding success in the absence of intense competition from their conspecifics, and numbers will rapidly bounce back.

I wander on as the path dips downhill and the wind drops. It is cathedral-quiet now. It's a depth of peace

that's just not attainable where I live in the busy south-
east of England. Yet I wonder, as I walk, whether even
here there are too many people around for there to be
wildcats. It's late autumn, but in summer I can imagine
these wide paths along the forest edges attract a profusion
of walkers, cyclists, runners. The spider's web of smaller
paths that criss-cross the central one are probably
quieter, but perhaps there are too many of them. The
RSPB, which manages this forest for its wildlife, reports
that wildcats (or at least wildcat-like cats) are seen
occasionally. I have seen a terrific camera-trap image of
one captured in the forest from 2000, which to my
amateur eyes had as fine a wildcat tail as anyone could
hope for, but no one ever saw this particular cat with
their own eyes.

The forest does have pine martens, that other and
altogether more attainable iconic Highland predator.
These gorgeous mustelids have also been caught on
camera, climbing up to the famous osprey nest down
near Boat of Garten to the south-west. Early-rising fans
of EJ the osprey – tenant of the nest since 2003 – were
alarmed to see, via the webcam aimed at the nest, the
inquisitive face of a pine marten appear over the brim of
the nest before dawn one April morning in 2018. EJ, a
tough old bird if ever there was one, bristled and shouted
at the intruder and saw it off, even though she had no
eggs to protect. Pine martens are skilled climbers and
expert nest-robbers – the osprey pair at another famous
nest, by Loch Arkaig, also had an April visit in 2018 from
a marten and then in May lost their clutch to (presumably)
the same marten, despite conservationists trying to
discourage the predator by cutting side branches from

the nest tree and coating the trunk with a slippery compound to make climbing more difficult.

Pine martens are close to cat-sized, though lighter-weight with their lissom bodies and very long, thickly fluffy tails. With their chocolate-and-cream coloration and handsome heads, they are attractive animals and, unlike wildcats, are increasing in number and expanding in range after plummeting to a worryingly low ebb by the mid-twentieth century. For me as a child with a wildlife obsession, pine marten and wildcat were the same in terms of extreme rarity and unattainability. I knew I'd somehow have to get to the remote Highlands to see either, and even then I would stand almost no chance. Both of these mammals had once occurred commonly throughout Britain. Centuries of persecution and forest clearance had left just a few hundred martens, and probably the same number of wildcats, in remote regions of Scotland.

But now, pine martens are thriving. They have benefited from full protection and from reforestation. Their numbers have grown to some 4,000 in Scotland and they are spreading south through northern England. They have also been reintroduced in Wales, and Ireland has a growing population as well. Such is their success that there are now calls for a cull from some quarters (a depressingly predictable reaction to news of any predator daring to increase its population in this country), even though this is still a very rare animal, and presents no significant danger to any human interests. In fact, the opposite may be true. Fans of the marten point out that it could be capable of carrying out a cull of its own – of the grey squirrel. Martens are agile and skilful treetop hunters, though red squirrels can often escape them by retreating to outer branches too spindly to take the martens' weight. For the heftier grey squirrel, this is not so easy, and it's possible that the pine marten could prove a highly effective controller of our most notorious invasive mammal species.

To see a pine marten, all you need to do is rent a cottage somewhere around suitable habitat, and put some bait out in the garden. Wait until dark and, with luck, you'll see a humpy-backed and bushy-tailed shape come lolloping into view and help itself to the chicken wing, smear of peanut butter, jam sandwich or whatever else you've left out for it. Unlike wildcats, pine martens have no overriding horror of humans. They can even be persuaded to live and have their pups inside human-made nesting boxes nailed halfway up a suitable tree. However, I'm still extraordinarily unlikely to see a pine marten here in the forest on my daytime wanderings even though we know for sure that they are here.

I walk a slow 10km loop and see almost nothing at all – even the small birds are elusive today. It's discouraging. I still head out again at dusk to my favourite forest-edge spot, but to no avail. I decide on a change of tack for the next day.

Hours of light are short, so I'm walking in near-darkness down the road into Nethy Bridge. I usually take the riverside path but it's too dark for that. From the village centre I find the northbound arm of the Speyside Way, which leads through more open countryside on its way to Grantown-on-Spey. I've heard that this stretch of path holds wildcat potential, so I'm more hopeful than usual. I've soon left the outskirts of the village behind and I'm on a track that traverses a picturesque rural landscape with fields (some stubble, others rough pasture), hedgerows, pine copses and tree-lined streams. The sky is brightening, displaying lilac streaks against indigo. It looks like it'll be a fine morning.

Already this pathway is serving up more wildlife than the forest. I stop to admire a rook that's alighted on a fencepost up ahead, to bow and give its hard-edged caw. Beyond it, more rooks are lifting up from the stubble field, their diamond tails and long-fingered wings making them seem gangly and irregular, not like the neat, compact jackdaws in their two-by-twos. I'm reminded of bonfire ash blown into the sky, and feel a twinge of melancholy.

Onwards and things get wilder. No houses now for a mile or more. I'm following the course of the Spey but the path veers closer and further away – now and then I can't even see the river at all. I disturb some roe deer in a small patch of forest and somehow grab a photo of one of them in its full, panicked flight, the polished black hooves caught in mid-leap. Out onto a more open stretch again and there are suddenly redwings crash-landing in the sparse trees around me, calling incessantly. I wonder if they're fresh arrivals from their Arctic summering grounds. Their voices are thin, coldly grating – the sound of Siberia. At a glance they look like our song thrushes, but song thrushes have mild expressions and each of these redwings' faces wears a frown or glare, painted that way with dark and light stripes around the eye and across the cheek. I try to photograph them as they fly, catching black-streaked bellies and rusty armpits as their compact outlines flicker over. Then I shift my gaze lower, and freeze, because there is a cat on the path ahead of me.

It's a black cat. So the adrenaline jolt is over almost before it's begun, but I take a careful look at this cat nonetheless. It's a long way up ahead and facing away from me, walking or stalking away along the grassy track in a slow, deliberate manner. Through the binoculars I can see it is a big, stocky cat, broad-headed and blunt-tailed. I think about the Kellas cats of legend – wildcat-shaped but jet-black, once thought to be a variant or even a different species or subspecies of wildcat and said to be even bigger and more ferocious, but all known specimens are now confirmed to be wildcat–domestic hybrids (or occasionally truly melanistic wildcats). I take a couple of photos of this black cat, which are not very good because it's still early and the light is low, and because it's so far away. I decide to try to get closer, but some sixth sense warns the cat of my intentions. It glances over its shoulder and immediately bolts away through the hedgerow and into the trees on the river's side of the path.

Another not-wild cat, then, but one whose appearance and behaviour makes me wonder if it could be a hybrid – I would certainly not be surprised if I learned (somehow) that it was a feral cat, living wild and off its wits long-term. But then again I have known friendly domestic cats that become terrified of people the moment they step through their cat flaps into the outside world. The story of this cat is unknowable – and that's the frustration of nearly everything to do with Scottish wildcats. So I keep walking. I stop often to scan the furthest fields, but the light is growing as mid-morning approaches, and surely any self-respecting wildcat would not be out and about any more. Yet again, I look to the other wild animals that share the wildcat's world.

A noisy bugling overhead heralds the arrival of a huge skein of greylag geese. I watch them as they beat purposefully around in a sinking circle, finally landing in a far-off field among hundreds more that I'd not noticed until now. Among them are a few white shapes. Back home, the greylags would be feral birds, and any white individuals would be recent farmyard fence-hoppers – the domestic goose descends directly from greylags. But up here the greylags are truly wild and have come here from Iceland. And the white birds with them aren't even geese – they are whooper swans, but they have come here from the far north of Europe and Russia. Between them and the redwings, all these birds blown in from the high Arctic, the mood is truly wintry.

Eventually I come to a lone farmhouse with signs on its fences about the ecologically friendly farming methods used on these fields, including leaving wild patches to provide winter food for finches and buntings. Fittingly enough, there are various finches fossicking in the hedgerow, and also a bunting right opposite the most prominent sign. It's a male reed bunting in fresh winter plumage tones of chestnut and ash. He is much less boldly patterned (but, in my opinion, more beautiful) than he will be in summer when he'll sport a glossy black head with contrasting white collar and moustache. In fact, no change of actual feathers occurs to bring about this seasonal change; it is achieved gradually, through wear. The feathers are tipped with those chestnut and ash colours but, as the feather fringes wear away over time, more and more of the black (or white) lower part of each feather is revealed.

A far-off mew catches my attention. It's not a cat but a buzzard, describing low, slow circles in the multicoloured sky. I look around and spot four more doing the same thing. I imagine them queuing up to board the first thermal of the day – a free ride that will swirl them up to a spectacular aerial viewpoint. From here, they'll drift for miles, soaring on flat wings with minimal effort, and scouring the ground below for anything that resembles breakfast. Another raptorial shape carves the vista in half – a mighty female sparrowhawk who has no time for thermalling and is dashing at breakneck speed towards a wooded hillside just beyond the farmhouse.

It's not too long before the path winds right down to the river's shore. The Spey here seems deep and slow, less frantic than it is further south. I pause to look downriver, wondering if I'll find a dipper or a kingfisher. I don't, but there is a little group of goosanders, which hastily paddle away around the first bend of the river as soon as they notice me watching them. Long-tailed tits approach through the riverside trees, conversing non-stop with soft ticking and purring notes as they navigate a complex twig-to-twig course. They bridge the gaps with bouncing flight, their exaggerated tail feathers dancing.

It's a long hike back, and the best part of the day is over now – at least until dusk, and at least as far as wildcats are concerned. I'm tired but the falling light

drives me out again that evening. I walk slowly, I sit and
wait, I hope and then I return. It's the same story the
next night, and each night after that, and the same in the
mornings too. Walking in the forest and the fields beyond
still soothes me, and the other wildlife I encounter still
enchants me, but I no longer really believe that I might
meet a wildcat someday.

CHAPTER FIVE

A Bloody History

Conflict is everywhere in nature: predators trying to kill prey; prey trying to kill predators by escaping and thus depriving them of their food; competitors doing battle over resources; parasites and thieves, mimics and cheats. Every individual animal is struggling to attain advantages that will allow it to survive and breed. Richard Dawkins says that all behaviour is driven by the selfishness of our genes and their blind need to exist into perpetuity. Through all this conflict, balance of a kind is the usual result because there is interdependence as well as conflict. But when one of the animals involved in a conflict of interests is *Homo sapiens*, the outcome tends to be the same every time.

In chapter 1 we looked at the wildcat's history in Britain. When the last ice age ended some 12,000 years ago and Britain was gradually freed from its permafrost prison, many mammals that had retreated to what is now the mainland were able to return. They crossed via what was then dry (well, perhaps damp) land, connecting eastern England to north-west Europe. Among these recolonisers were wildcats – and also humans.

The modern geological epoch, the Holocene, began about 11,700 years ago. This time also marked the start of the Mesolithic period – the last era in which human beings were pursuing a truly hunter-gatherer lifestyle. They collected plant matter and used simple (but quickly more complex) tools to hunt animals. In Britain, which quickly became fully forested as the ice retreated, those animals were the likes of red and roe deer, wild boars, and some that are now extinct such as the mighty Irish elk and the native European wild ox, the aurochs. Smaller furry quarry might have included mountain hares in some areas, but this was before rabbits or brown hares were living in the British Isles. People would also have speared fish and caught birds when they could. Communities were not necessarily permanently settled in one place, but neither were they constantly nomadic and many long-term Mesolithic dwellings have been discovered.

At this time in history, there would have been no appreciable conflict between humans and wildcats. The wildcats didn't compete with human interests in any obvious way. Fossil remains show that wildcats were present throughout England and Wales and they spread north into Scotland as the climate warmed up. There is also evidence that they reached several island groups, such as parts of the Inner Hebrides and the Isle of Man, and they almost certainly got there by swimming rather than being transported by humans.

It is also quite likely that wildcats reached Ireland. Anecdotal reports from the last two hundred years are quite convincing but they are not backed up with any actual specimens – there is no real evidence that wildcats lived in Ireland any more recently than the middle of the last millennium, despite exciting nineteenth-century accounts of fearsome tiger-striped felines that (like their counterparts in Great Britain) could attack and kill almost anything, including people.

However, enough bone remains have been found to strongly suggest that wildcats were present at least from 10,000 to 3,000 years ago in Ireland across a broad swathe of the southern half of the country. The earliest of these carbon-dated bones precede the arrival of the domestic cat in Ireland. However, there is a possibility that the bones found were from wildcat corpses brought over from the mainland by people looking to trade the skins. Cats are mentioned in a variety of old Irish Gaelic-language place names – for example, there are several places in Ireland called Lisnagat (*Lios nag Cat* – the fort of the wild cats) and Carrickacat (*Carraig an Chait* – the rock of the wild cat). However, 'cat' in Ireland can also mean pine marten, so not too much store can be set by such names.

That wildcats ranged throughout Great Britain, though, is in no doubt. And there is every reason to suppose that human and wildcat coexisted in relative peace up until the Neolithic era began, about 4300 BC. Then everything began to change.

The hunter-gatherer way of life had its limits. When human numbers expanded to the point that hunted and foraged food was no longer enough, the human talent for innovation came to the fore, and communities began to develop new ways to feed themselves: through keeping and breeding their own animals, and growing their own plants. This meant an end to nomadism, and establishing permanent settlements close to the places where crops were grown and livestock was kept. To create open space for farming, people cleared tracts of forest, felling trees with stone axes. They also used fire and ring-barking (cutting a complete circle of bark from a tree trunk, which would kill the tree) to remove trees. Non-native animals were brought to Britain from Europe and further afield – early domesticated forms of chickens, cows, pigs and sheep began to arrive. The wildwood that cloaked Great Britain started to shrink in the wake of advancing and developing human civilisation.

Managing resources that they considered 'theirs', and which they worked hard to nurture, brought humans into conflict with nature in a new and very destructive way. There was no longer such a need to hunt wildlife in order to eat it, but instead wild animals (mostly *less edible* wild animals) had to be killed because they might otherwise eat the humans' crops and livestock. Wildcats would kill poultry, so they (along with pine martens, foxes and other predators) were regarded as enemies, vermin. There was no sensible reason for a Neolithic community not to kill every wildcat they could. By this time there were domesticated dogs in Britain, and they would have been used to kill wildcats and other 'pests'.

Deforestation and growing human settlements were focused around certain key areas where the land was very productive. These were primarily the Upper Thames Valley, Wessex, Essex, Yorkshire, around the Wash in East Anglia, and also Anglesey, the Orkneys, eastern mainland Scotland and the River Boyne in Ireland. The rate of spread of the human population was still modest by modern standards. However, things accelerated with the advent of the Bronze Age in about 2000 BC, and even more so with the Iron Age beginning in about 750 BC. Iron ploughs and axes speeded up all of the agricultural processes dramatically, and allowed much further and faster spread of humankind throughout Britain.

Meanwhile, the African wildcat was becoming domesticated, a process that probably began in Egypt around 3500 BC. Its value around the house was down to its rodent-catching abilities, but Egyptians were famously beguiled by its more aesthetic qualities. They kept the animals in their homes rather than outside, lavished love upon them, mourned their deaths, used their likeness as a basis for various deities, and celebrated them in art of all kinds. It's not a stretch to say they were the first crazy cat-people in history. And as Egypt made contact with other cultures of the time, so the domesticated cat spread across Europe. It wasn't revered everywhere as profoundly as it was in Egypt, but its usefulness as a guardian of food stores was apparent. The Romans took to the domestic cat, and they brought it to Britain with them when they began to arrive here, soon after AD 40. Their arrival also marks the end of the Iron Age and of prehistoric times.

Through the next millennium, the growing human population continued to modify the British environment and by AD 1000 wildwood covered only 15 per cent of the land in England. It continued to shrink, and pressure on Britain's wild predators was stepped up more and more. The Eurasian lynx was already extirpated, gone sometime around AD 400, and by AD 1000 we had also lost the brown bear. The wolf survived longest of all the large carnivores but by the late seventeenth century it too was lost from Great Britain and soon afterwards from Ireland. The Tudor 'Vermin Acts' – laws introduced in the sixteenth century – dictated that gamekeepers, landowners and farmers could claim a bounty for each of the 'vermin' animals they killed. The list of species included was extensive: obviously all predatory mammals, owls and raptors were there, but also many others that surely must have seemed as blameless then as they do now – the likes of hedgehogs, woodpeckers and kingfishers.

For wildcats, the bounty was set at one penny – awarded when the killer presented the wildcat's head to the local church warden. Records kept by parishes of the payments given show that some 1,100 wildcats were recorded as killed through the seventeenth and eighteenth centuries. Given that the surviving records covered just 14 per cent of parishes in England and Wales, we can extrapolate that at least 8,000 wildcats were killed in total under the Vermin Acts – about 5,000 in the seventeenth century and 3,000 in the eighteenth. (You can read more of this in Roger Lovegrove's 2007 book *Silent Fields*.) The difference between these two figures almost certainly reflects a general decline in the wildcat population rather than reduced effort from the landowners.

The Acts inevitably led to wholesale extermination of wildlife throughout the countryside, and many species suffered catastrophic declines. Already by the start of the seventeenth century it was no longer true to say that wildcats were widespread in Britain, and things went rapidly downhill from there. Persecution continued apace through the eighteenth and nineteenth centuries, and this – coupled with the almost complete loss of forest – was too much for many species. Several species of birds of prey were extirpated, and numbers of mammals like pine martens, polecats and wildcats were devastated. Wildcats were soon gone completely from whole regions, first from southern counties and then northern. The Vermin Acts were repealed in 1863 but anti-predator attitudes remained as intractable as ever. The last known Welsh wildcat was killed in 1862 or thereabouts, and the last few were lost from northern England before the start of the twentieth century. They were also extirpated from the southern Scottish lowlands.

That wildcats managed to hang on in Scotland's uplands is in no way due to kinder human attitudes, but down to a more difficult terrain and a smaller human population. Landowners still killed them at every opportunity, along with raptors, owls, corvids, pine martens, polecats, stoats, weasels and foxes. As more and more Highland landowners began to use their land as game estates for grouse-shooting, the pressure increased. Gamekeeping as a career path took off in spectacular style, and predator control was one of the most important aspects of the job. With the advent of shotguns and the invention of various efficient if horribly cruel trapping methods, killing predators became easier and easier and the predators became more and more scarce. The leg-holding gin trap was widely used to trap mammals, and many accounts exist of wildcats mutilating themselves to escape from its steel jaws. Throughout the nineteenth century and into the early twentieth, war was waged on all wild animals equipped with claws and fangs.

Some Scottish game estates kept detailed records of the 'vermin' they killed. The Glengarry estate in north-east Scotland was one such: their records from 1837 to 1840 show 198 wildcats were killed on the land in that period, along with 246 pine martens, 106 polecats, and 301 stoats and weasels. The estate also recorded 78 'house cats going wild', a big enough number that hybridisation between feral and wildcats was probably already a significantly frequent occurrence. In later years, some other estates killed very large numbers of roaming formerly domestic cats; for example, 4,327 of them on the Buccleuch estates in Dumfriesshire between 1894 and 1900. Gamekeepers at this time, of course, had no qualms about destroying free-ranging cats of any kind, and their efforts continued through most of the twentieth century. The wildcat finally received legal protection in 1988, as the government acknowledged how rare it had become, but feral cats were not protected. It took several more years to raise awareness of how to distinguish the Scottish wildcat from ferals and hybrids, during which time many more Scottish wildcats were illegally killed.

The pattern through the last few centuries has not been of uniform habitat loss and wanton destruction, though. If it had been, the Scottish wildcat probably would have gone the same way as the lynx, wolf and bear. Game-shooting estates were not friendly places for predators, but on deer-stalking country a considerable amount of tree-planting took place, beginning as early as 1750. This provided Scottish wildcats with new habitat that suited their needs better than the much more open landscape that it replaced, and helped them to survive better, provided they kept out of the way of gamekeepers. Another helping hand came when wartime duties from 1915 took many gamekeepers away from their work. This gave breathing space to all kinds of wildlife, and the wildcat population in Scotland appears to have begun to recover and spread back to former haunts over this time. However, later in the twentieth century persecution resumed at an even higher rate.

Although wildcats were never considered to be worth hunting for food, they did yield an attractively patterned (if coarse-textured) pelt, and some of their body parts were used in traditional medicine. Their fierce natures also made for an exciting chase and kill and accordingly wildcats were hunted for sport as well as being controlled as pests. The same was done to foxes, pine martens, otters and other predators, as well as hares, after the Norman Conquest in 1066. William the Conqueror was a keen hunter and established several royal hunting forests for his own use. These were expanded and the laws governing them made more elaborate by subsequent monarchs. Hunters used dogs to chase down and kill quarry. When royal parties were out hunting the king's deer, non-target animals like wildcats and foxes ('beasts of the chase') would also be pursued and killed whenever encountered. For ordinary people, taking actual game (deer and boar) in the king's forests was outlawed, but some individuals were allowed onto the land to hunt wildcats and the like in the forests by licence.

It is from hunters that we hear some of the most dramatic stories of what the wildcat can do in self-defence mode. The *Glasgow Herald* in 1875 reported a dog blinded by a wildcat's furious swipe, while *The Times* in 1837 carried a report of a wildcat that had killed both of the dogs set upon it. However, more usually the dogs would drive the wildcat up into a tree, which could save their lives if they were then able to cross to another and stay beyond reach. A determined wildcat could often hold off one dog on the ground, though, and sometimes hunters had enough respect for the cat's bravery that they would allow it to escape. On the other hand, a cornered wildcat would not hesitate to attack its human tormentor and inflict some terrible wounds. A legend from fifteenth-century Yorkshire tells of a nobleman attacked by a wildcat in the grounds of St Peter's Church. He managed to kill the animal by crushing it against the church wall, but he then died of his injuries and his body was found in the church porch the following morning.

As well as being killed to protect livestock and game and for sport, to some extent wildcats were also killed for the use that could be made of their bodies. In northern latitudes, fur-bearing animals were and are valuable resources, as we humans had only recently (in evolutionary terms) left the African savanna and our bodies had not had time to adapt to the new and markedly different climates in which we found ourselves. Warm garments were more and more necessary the further north humans migrated, and many different kinds of mammals paid the price for our need for fur. Soft, warm and beautiful fur belonging to the likes of the pine marten, red squirrel or stoat (particularly in its white winter pelage) was used as a mark of high status in medieval times, and the dense pelts of beavers, otters and seals were prized for their robust qualities, but the fur of all animals had value, including that of the wildcat.

Wildcat fur is on the coarse side but is said to be very warm, and was considered of better quality and to offer higher weather-resistance than that of domestic cats, which were reared and killed for their fur at this time as well as for their rodent-hunting skills. Cat pelts commanded a reasonable price, and were often used as lining material. Through the fourteenth and fifteenth centuries new laws were passed, limiting the kinds of fur that people of different social class and income were permitted to wear, partly as an identifier of rank but also in a bid to protect dwindling populations of some of the fur-bearing animals. However, wildcats remained on the 'free-for-all' list. By the start of the twentieth century, fur use had declined considerably in Britain as it became more and more difficult to obtain pelts and as new warm fabrics were devised.

Sometimes wildcat fur was chosen not for its practical qualities but because of what it represented. Several clans take the wildcat as their emblem, in respect for its power, ferocity and proudly untameable nature, and clansmen once wore sporrans made of wildcat pelts. Historically certain Pictish and Celtic tribes sometimes wore wildcat skins, often incorporating the animal's head with open jaws.

We (at least we in the West) baulk at the idea of eating cat flesh today but we didn't always have the luxury of choice. The first British settlers, pursuing a hunter-gatherer lifestyle, would have eaten every animal they could kill, including wildcats. There is much archaeological evidence for this in the form of bones found around settlements that show cut-marks indicative of butchery. Through medieval times and beyond there was less eating of cats wild and tame, as it was becoming easier and easier to keep more delicious animals for the purpose of making meals of them. However, a few cultures did keep a taste for wildcat flesh – there are, for example, plenty of recipes for cat-based dishes from France and Spain in the nineteenth century.

It was well known that wildcats could not be kept as pets, but that didn't stop people being curious to see living wildcats close up, especially as they became rarer and rarer. In the late nineteenth century, cities in Britain hosted the first cat shows, intended to show off the diversity of domestic cat breeds at the time, and at the first of these – at Crystal Palace in July 1871 – a wild-caught Scottish wildcat was among the 170 or so exhibits. The show attracted far more attention than expected, with 20,000 visitors through the doors. The wildcat, which had been snared on the Duke of Sutherland's estates, had lost its forepaw to the trap that caught it. It must have been in great pain and terrified; the judges described it as a 'savage varmint'. What became of the poor animal after the show is not known and we can only hope it didn't suffer too much more.

As well as a source of meat, fur and entertainment, the wildcat was also used as a living medicine cabinet, as were nearly all other wild animals before anything much was known about how to treat various health conditions properly. In the Middle Ages body parts from a bewildering range of animals, prepared in an equally bewildering variety of ways, were held up as surefire cures for ailments as diverse as epilepsy, cancer, gout, haemorrhoids, leprosy and impotence. In *The Compleat English Physician* by William Salmon (published in 1693), we learn that blood taken from a wildcat's tail and mixed with salt, ground-up ox horn and ground-up human skull (in precise proportions) will cure something called the 'falling sickness', while both the fat and the flesh were effective treatments for gout.

The campaign of destruction waged against wildcats had brought their numbers down to critically low levels by the late twentieth century. Only a handful of the victims found their way – as taxidermied specimens – into museums and private collections. Examples can be seen at the National Museum of Scotland in Edinburgh. Historical Scottish wildcat specimens have proved of crucial value for modern wildcat conservation, as they offer a source of 'pure' DNA with little if any dilution from hybridisation with domestic cats.

To those of us today who spend our free time watching wildlife, the wildcat is a magical beast, almost sacred in its extreme rarity and elusive character – in short, something we would all love to see although we know we probably never will. Perhaps this is an overly romanticised image of what is, after all, just a wild animal doing what it can to survive. But the public image of the wildcat a couple of centuries ago was almost diametrically opposed to this, and the views of the time make for quite shocking reading. In his compendium of our wildlife (*British Zoology*, 1776), the naturalist Thomas Pennant describes it thus:

> *The cat in its savage state is three or four times as large as the house-cat. It may be called the British tiger: it is the fiercest and most destructive beast we have; making dreadful havoc among our poultry, lambs and kids.*

Pennant's name has stuck even today, though now our Scottish wildcat is nicknamed the Highland tiger. The name is a compliment now, acknowledging the cat's beauty and rarity as well as its fierceness, but people of the time despised and feared wildcats, and saw them as destructive, dangerous, and wholly without redeeming traits. It took centuries and the almost total loss of wildcats from our countryside for attitudes to change, but now we are fighting to turn back the clock and rescue – if not recreate – our magnificent native felid. Persecution is not so serious a problem now – an even more intractable issue has taken its place.

Speyside and Beyond, 2015

When the opportunity crops up to join an organised wildlife tour in the Speyside region in September 2015, I leap at it. Many birding and wildlife-watching tour companies cover this area with a tour or two each year, but I am travelling with Speyside specialists – Heatherlea, whose headquarters at the Mountview Hotel are just a short(ish) walk from my previous temporary home up here at Dell Lodge. The tour I am doing promises an array of classic Highland wildlife. It doesn't promise wildcats, but I wouldn't trust such a promise anyway. We'll be in Speyside at the start and end of the week, with a couple of days out on Lewis and Harris in the Outer Hebrides in the middle.

I arrive at about lunchtime, and a few hours later I tear myself away from my hotel bed (the overnight coach isn't getting any more bearable) to meet the rest of the group. There are a dozen fellow wildlife-watchers here and they are all charming; likewise our two guides. I'm not used to communal wildlife-watching – I'd rather be out with just one or two friends or on my own – but then again many eyes can make for many sightings. With all these people, surely the chances of seeing something as shy as a wildcat are minimal, but then the itinerary says we'll be seeking capercaillies and they are similarly shy.

We kick off day one with a pre-breakfast drive out into the hills north of here, and park up on a miniscule road clinging to a steep slope. Uphill is heather moor;

the other side slopes down to pasture with a few stands
of trees. No humans around – the only sign of human
presence in fact is the fence-line below and it's this
where our attention is focused. Moving around on the
ground at the foot of the fence is a big, plump-bodied
grey bird: a female black grouse, or greyhen as they
once were known (the males were blackcocks). There
are four kinds of grouse in Scotland and they each live
in slightly different habitat types. The capercaillie only
lives in large pine forests in the lowlands. The black
grouse prefers hillier, grassier ground with copses, young
plantations or the edges of forest. The red grouse is
found on higher ground again, on the open heather
moorland – indeed, we can hear them calling from the
uphill side of the road. And right on the mountain
peaks, where even the heather dwindles away and the
landscape turns into tundra, is where you'll find number
four: the ptarmigan.

Of the four, only red grouse are easy to see (and that's
largely because there are so many grouse moors up here,
and grouse moors are managed in such a way that
artificially high numbers of red grouse are kept alive for
the shooters to kill). Ptarmigans are not exactly elusive
but you have to endure a hell of a climb to reach their
altitude. Capercaillies, while the biggest of the lot, are
shy to the point of invisibility, and extremely rare as well.
Black grouse, after a torrid time of it a couple of decades
ago, are recovering in number but are still far from easy
to find, so everyone is happy to kick-start the week with
reasonably good views of one. Then one becomes two as
a second greyhen flies in to join the first, then three as a
fine young male alights on a fencepost and sits poised

there, showing off his lyre-tail and the scarlet comb above his eye.

It's a cold morning and cloudy – the sun-rising sky an abstract tapestry of improbable colour that quickly fades to leaden as the hour advances. We mill about at the roadside a little longer. Fieldfares, new arrivals from the Arctic, are heading over in dribs and drabs – sturdy arrows of birds. Hearing their peevish, clucking calls for the first time in months is a potent reminder of winter on the way, and jars with the bright 'vit-vit' calls of the lingering swallows that wait on overhead wires. I scan the scene methodically through binoculars, paying particular attention to the landscape lines – the walls, the fence, the general rolling lie of the land. Quiet and remote, this vista looks like it has wildcat potential, and the view stretches far enough that (perhaps) a wildcat could show itself in the distance without noticing our little group on the hillside. But all I find is one or two distant roe deer, grazing along the treeline edges and looking like they're thinking about returning to the safety of the woods to pass the daylight hours.

Our morning is spent tramping around a pine forest that I have never visited before, on a ridge behind Grantown-on-Spey. We seek capercaillies and we find their big, three-lobed tracks in the mud – the guides also point out some caper poo, which resembles the ash left by a fat cigar. We walk to a clearing with a long view, where capercaillies might fly from left to right, but none do – the only aerial animal in evidence is the dreaded Highland midge, which goes to town on the bare skin of all of those foolish enough not to have slapped on a layer of insect repellent (that would be me).

Later on we drive another way and we're soon heading down another tiny road along the course of the Findhorn River, into hills that grow into mountains. At our lunchtime stop-off, a juvenile white-tailed eagle glides high overhead, dwarfing the buzzard that is pointlessly mobbing it. We follow the road to its end and walk a little way along footpaths towards the biggest mountains. I pause on a bridge over the river – now more of a rapid and noisy stream – to watch a dipper dancing from stone to stone. One of our two guides, Jonny, goes off in search of mountain hares on the lower slopes but has no luck. I chat to the other, Phil, about wildcats and the impossibility of seeing them.

The following morning, we strike out westwards. It's reassuring to know that, even on a long drive with a time limit in mind (we're boarding a ferry from Skye this afternoon), the guides will stop if anything of interest is seen on the way. Accordingly, I'm on full wildcat alert, glued to the window as we traverse this lovely landscape. There's no straight-line route for us – up here we are funnelled along the valleys where they fall, and the hills and mountains tower over our little bus. Then we're crossing the broad bridge to Skye.

Wildcats once lived on Skye, according to eighteenth-century naturalist Thomas Pennant. They may have reached the island by ancient land bridges when sea levels were lower or they could have made the relatively short swim from the mainland. Recolonisation now, though, seems highly unlikely. I have to set aside my wildcat hopes now, and for the next two days, as we explore Skye and then head out to the Western Isles. We cross Skye slowly, with frequent birding stops, and then it's time to head for Uig in the far north to board the ferry to Tarbert on Harris.

I love a ferry crossing even when it's taking me away from wildcat country. The prospect of exploring Harris and Lewis (the respective names for the south and north of this one long, frilly-shored land mass) is exciting – the main draw is the chance of seeing a golden eagle, to go with the white-tailed that we saw yesterday. But this is another level of wild Scotland too. We drive north from Tarbert and we are soon in what's almost a moonscape. Compared to the richness of the Speyside forests, this wide-open, rugged, rolling and rocky landscape seems brutal, bleak. But then we park up on a clifftop and look out over the most beautiful white-sand bay, the shallows marbled cream and silver, and it's lovely. A moment later a raven flies past at eye level, and in its colossal bill it's carrying a terrible trophy: the severed head of a rabbit. And we're back to brutal again.

At some point, up on this cliff, Phil finds a golden eagle in the expanse of sky. He lines up his telescope on the distant wheeling speck. I take my place in the queue to look. It's moving away – it's a long-winged shape and I can't make out anything more than that. I step away quickly so the rest of the queue can see it before it's gone. I can't pretend it's a deeply satisfying sighting, but then it's how I've come to imagine golden eagles – my fortune at finding them has been nearly as bad as my wildcat luck. A faraway and fast-vanishing speck seems about right. Maybe we'll get a better look at another one later – we've a full day and a morning still to spend here on the island.

We pass virtually no signs of humanity on the drive to the northern tip of Lewis, but Stornoway is biggish and busyish – it feels like a proper town. Some of us even consider seeking out a pub for after dinner, but in the end I'm too tired and I head for bed early.

Our full day on Lewis brings a string of moments to remember. Another goldie, which we find as it's already heading away and, again, I manage a glimpse of a speck through the scope. A merlin that streaks past us as we're climbing out of the bus at a vast and empty golden beach. Standing on the clifftop at the Butt of Lewis and finding, far out to sea, a twisting and turning flock of sooty shearwaters dipping over the wave crests. Jonny, a former reindeer herder, using his special ululating call to confuse a passing flock of golden plovers into settling almost at our feet. A late-afternoon snack by a river with gulls hovering, eager for thrown scraps of cake. I'm too tired for the pub again this evening.

On the last day, we're sailing out of Stornoway and travelling a longer route back to the mainland – to Ullapool. Our guides call us together at the ferry port and tell us it's important that we try to claim seats at the very front of the big ferry we're about to board. This crossing could be good for storm petrels, but sitting at the front will be the only way we stand a chance of seeing one, so we have to hurry. My ears prick up at this. I'm the youngest person in our group and I also did the Great North Run a week ago – this is my mission. As soon as we're let on board, I and the second-youngest person in our group start jogging together along the corridors, weaving around various slowcoaches with wheelie cases. It's not, in the end, a tricky challenge at all because the boat's quiet and we seem to be the only people bothered about sitting at the front, but there's still a sense of personal victory as I choose a place in the front row of the cinema-style seats. Soon the rest of the group are with us too, and we're all packed into the small, pointy space, staring out of the big windows as the ferry draws out of the port.

It's soon clear that seeing anything at all will be difficult. The ferry is so big and high, and storm petrels are so tiny. It's hard enough spotting bigger seabirds – bonxies and gannets, auks and shags. I miss the first petrel completely, but the second I just about lock onto for a moment. It's like a tiny scrap of paper, a black-and-white moth against that huge rolling mass of sea, veering haphazardly into view then gone almost immediately.

The light's fading when we reach Ullapool. I'm falling asleep through the black miles back to Speyside. A dozen wildcats could have crossed our path, turned their shining eyes into our headlight beam, and I would have missed them all.

The next day takes us up to the Moray coastline to search for seashore and farmland birds. I'm back on wildcat watch, scanning the fields, the hedge-lines, the edges of forest – the usual deal and the usual result. But although there are no wildcats, there is an osprey resting on a post in Findhorn Bay, and there are swarms of waders dropping down to the shoreline to feed, and there are thrushes and buntings and finches in the fields. In the dying hours of the day we stop off in Abernethy Forest and find ourselves surrounded by calling crested tits. They pose in the small pines that fringe the car park, showing off their punky hairstyles. Overhead, redpolls and crossbills fly fast across the clearing. We're ostensibly looking for capercaillies here – the glaring gap in our target bird list, and one that we'll be trying in earnest to find tomorrow.

Our last day dawns. We've bonded as a group and conversation flows, but we're also back on caper-watch so we have to shut up. The guides remind us to step softly, to murmur quietly and to stick together, but we're a sizeable group and we still seem to make a hell of a racket as we walk down this broad forest path. And it still strikes me as faintly absurd that we're genuinely expecting we can turn up like this and stand a chance of seeing one of Britain's shyest and most hypersensitive wild birds. This is the fourth and final forest we're visiting in search of capers, so there's some tension – one or two of the people in this group have never seen a caper and desperately want to. The signs don't seem great. But then there's a small commotion in the treetops ahead, and those of us who look up in time manage to catch a glimpse of an alarmed female capercaillie as she dives out of view into the dense forest. I'm not one of them – I was looking down into the lower slopes, my mind (as ever) on wildcats. But we're more fortunate than I could have dared hope, and we see a second capercaillie in almost exactly the same way just five minutes further along the path, and this time everyone manages a look.

The relief is tangible as we return to the bus and head off for our final destination. I'm excited about this last stop – a walk up the stony flank of Cairn Gorm itself, one of the biggest peaks around, in search of ptarmigans. We park up at the point where the funicular railway carries reluctant walkers to the Ptarmigan Restaurant, close to the summit. Some members of our group have decided to take this route; the rest of us strike out on foot.

I am not very good at walking slowly up steep hills – it seems to be more tiring than heading up at top speed – so I have soon left the rest behind. I'm getting lost in thought as I look down, watching my step on the scarily uneven ground, when a call from Phil pulls me back to reality. He's pointing – I follow his gaze and there is a little group of ptarmigans picking their way across the boulder field. They are still in their mostly grey breeding plumage and incredibly camouflaged – if they stop moving they vanish, airbrushed into the landscape. The others join me and we watch the fat little grouse scouring the rocks for scraps of food. Then abruptly they have had enough of this spot and they take to the air, whirring past us and down the slope. In flight, they reveal pure white wings, and as they fall away from us through the air they become snowflakes, quickly swallowed up by this sprawling alien landscape.

There'll be no wildcats up here, I know that, but there is the chance of another mammal I'd like to see. The mountain hare lives on these slopes, and when Phil offers to lead some of us the long way down, I go along with him because this seems to offer more chance of meeting one. So we zigzag our way across an open boulder field, stepping over and between the slabs of lichen-cloaked rock and pausing often to look for a running or a hiding hare. It's a hard walk, a real challenge for me as I don't have as much faith in my balance as I should, and it's fruitless. But then as we near the bottom and spot the rest of our group, already back near car-park level, I can see that they've got something interesting in Jonny's scope, and as I look, one of the group holds his hands pointing up at the sides of his head, making a pair of hare ears. I feel a spark of excitement and complete the rest of the descent with reckless haste. There indeed in the scope view is a mountain hare, sitting stolidly in the shade of an overhanging rock. It, like the ptarmigans, is in its summer clothes of grey and white. It is sitting very still with its ears pressed down to its back and its eyes half-closed, emphasising the blunt blockiness of its face – frankly, it looks a little dorky, but I don't care. Jonny offers to lead me a little closer to its position and I accept eagerly, hurrying ahead of him. From the closer vantage point we manage to reach, the hare is bigger in my viewfinder but still ever so dorky-looking. I don't care. I take a string of photos of its adorable, dorky face. And wait a little longer in case it decides to move, but it doesn't. I don't mind.

On the last ride back to the hotel, I talk to the guides about wildcats. Jonny has seen them, Phil has not. Jonny has a few years on Phil, to be fair, but they both tell me that encounters are vanishingly rare even for people who live and work outdoors in the best possible areas.

I head home the next day with a head full of memories of all I've seen, with the capercaillie and the golden eagle occupying joint-second place on the podium of iconic Scottish wildlife, but the top spot remaining, as ever, unclaimed.

CHAPTER SIX

Of Cats Tame and Wild

A distressingly high proportion of British wildlife is declining and under threat. The RSPB's *State of Nature* reports bring together data on thousands of plant and animal species from across the UK, and draw a series of depressing conclusions: 56 per cent of species are declining; we are down about 44 million breeding birds since the 1960s; and wildlife abundance overall is falling, with invertebrates affected most severely – for example, 72 per cent of our butterfly species are in decline. The Scottish wildcat is not alone in facing its darkest hour.

The causes of this unfolding wildlife apocalypse are varied, but most significant of all are changes in the way we manage land, particularly for agriculture, since the 1960s. Intensive agricultural techniques to increase crop yields have taken away wildlife habitat on a grand scale, and pesticide treatments of the crop fields themselves make them inhospitable – little better than deserts. Habitat loss is also ongoing as our towns and industrial zones expand.

Another important factor is climate change, which is steadily pushing our northern and upland species – like the dotterel and the mountain ringlet butterfly – further north and further uphill, to the point where they are running out of space and being literally pushed off the map. Climate change is also allowing more southerly species to spread to our shores. This means exciting times for birdwatchers and entomologists, noting increasing arrivals of birds like the great white egret and insects like the southern migrant hawker dragonfly. These species are spreading north from the mainland because they can – it's easy to migrate to new lands when you have wings. But at the southern edge of their range, climate change is making survival more difficult for them. Their overall distribution is being slowly pushed northwards.

Other causes of wildlife declines include competition or other conflict with non-native, invasive species – the red squirrel's retreat in the face of the North American grey squirrel's northwards march is perhaps the best-known example of this. A few species are suffering intense, deliberate persecution – most notably the hen harrier, a beleaguered bird of prey with a fatal preference for nesting on grouse moors. Urban wildlife is suffering from our tendency to concrete over our gardens and block up all points where birds might nest in our buildings. The lack of large native predators in our uplands means rampant deer populations that damage plant communities. And for some declining migrant species, the problems they face are occurring far from on our own soil – for example, migrant birds that travel to southern Africa for winter are having to contend with desertification in the Sahel region south of the Sahara – a vital refuelling zone after the desert crossing.

For the Scottish wildcat, several of these factors apply, including habitat loss, climate change, and, sadly, deliberate persecution – this last factor extremely prominent in its history, as we saw in the previous chapter. The key problem it faces today, though, is unique: no other species in Britain is under threat of eradication through hybridisation with its own domestic cousin. Our beloved pet cats and their feral, free-ranging descendants are bringing about the destruction of our own native, famously ferocious wildcat, and it is an incredibly difficult problem for conservationists to tackle.

According to pet-ownership data collected by governmental surveys and analysed by the People's Dispensary for Sick Animals (PDSA), as of 2018 there are an estimated 11.1 million pet cats in the UK in total (although other surveys place the number lower, at somewhere between 7 and 8 million), with 25 per cent of adults having at least one cat. In Scotland, there are in the region of 880,000 pet cats. Estimating the number of feral cats is much more difficult, but it's likely to be about a million in the UK overall, with as many as 40 per cent of those in Scotland. Cats are our most popular pets, well ahead of dogs in this particular rivalry (there are only 8.9 million pet dogs in the UK according to the PDSA's figures). Dogs that stray from home tend to be caught quickly, and numbers of truly long-term feral dogs in Britain are very low.

Like dogs and most other domesticated animals, cats have been selectively bred over many, many generations to produce different strains or breeds with different characteristics. Because cats aren't really much use for actual work (except rodent control), their various breeds have been created to satisfy our aesthetic rather than practical needs. Random gene mutation has thrown up an array of coat colours, from red and blue (really ginger and grey) and jet-black to pure white, tortoiseshell and colour-pointed (Siamese-patterned). Then there are variations in hair length and texture, and cats with a shortened or missing tail, oversized or folded ears, or shortened legs. Body shape ranges from thickset and stocky – or 'cobby' as cat breeders would say – to the very slim, attenuated and whip-tailed oriental short-hair breeds. Some traits are controlled by a single gene and others (such as head and body shape) by many. Cats have escaped most of the extreme and health-compromising developments that affect many dog breeds, but some Persians are so short-faced that breathing is impaired, and the stumpy-legged 'munchkin' breed cannot move as freely as a normally proportioned cat. More controversial is the increasingly popular practice of introducing genes from wildcat species into domestic cat bloodlines, creating beautiful but often behaviourally problematic hybrid breeds. Even so, most pet cats are 'moggies', not pure-breeds, and many of them have the same short hair, medium build, and brown tabby pattern as their wild ancestors.

In one key respect, cats are not like other pets. In the UK at least, we have a strong tradition of allowing our cats free access to the outside world. We don't do this for other pets or indeed for livestock – we contain and control our animals. The only real parallel that I can think of is the domestic pigeon. Racing pigeons and some fancy breeds kept for their skill in flight are allowed considerable liberty, and it is no coincidence that urban and seaside areas of the UK have a vast population of freely breeding feral pigeons, since not every pigeon set loose will return to its loft. Their spread threatens the genetic integrity of their wild ancestor, the rock dove, in a similar way to how feral cats are erasing Scottish wildcats, but the important difference is that feral pigeons and rock doves are the same species. Scottish wildcats and feral domestic cats are not. They just happen (regrettably) to be closely related enough to interbreed.

Cats appeal to us as pets for myriad reasons. In general they are friendly but not needy. They have a relatively long lifespan, they are soft and cuddly, easily house-trained, quiet and entertaining. There's no doubt that another big part of their appeal, at least for some owners, is that they are low-maintenance. Fit a catflap, train your cat to do its business outdoors, and one of the biggest downsides of pet ownership is taken away at a stroke. Free-ranging cats are probably the easiest of all pets to keep and care for, but that ease comes at a cost. For a start, that business-doing has to happen somewhere – quite possibly in a neighbour's garden. There are also the hazards of outdoor living – accidents on the road, violent fighting with other cats, the risk of fatal encounters with dogs or even (for infirm or very young cats) foxes, or coming up against hostile humans.

The environmental impact of free-ranging pet cats is also cause for considerable concern: in a UK context a cuddly pet cat is (at least potentially) an alpha predator capable of hunting and killing a great variety of wild animals. Of course, not all of those 11.1 million pet cats in the UK are allowed outside access and, of those that are, not all are hunters. But there's still an enormous number of cats out there preying on wildlife. Compare this figure to some of our native, wild predators' numbers: the stoat's UK population is in the region of 450,000, and the weasel's is similar, while our fox population probably doesn't top 300,000. When it comes to birds of prey, of our commonest three raptor species, the sparrowhawk has a UK population of about 80,000, the buzzard's is approaching 200,000, and the kestrel population is about 100,000. Add these six common wild predators' numbers together and you still get barely more than 10 per cent of the total UK domestic cat population.

The Mammal Society's research on the impact of domestic cat predation on wildlife concludes that the UK's pet cats take 275 million prey items a year, of which 55 million are birds. The rest are mostly small mammals – mice, rats, voles and shrews – but also include many large insects as well as frogs, lizards, slow-worms and fish. Cats will also take other predators on occasion – I recently saw one trotting across a path on a nature reserve, holding the corpse of a stoat proudly aloft. That

275-million figure averages out at just under 25 prey items per cat per year.

Per cat per year, the predation rate is pretty low. Even if we assume only 10 per cent of all domestic cats are successful hunters, that's still only 247 prey items a year per cat – far fewer than are taken by wild predators, which almost invariably have to kill at least once every day. But where domestic cats and wild predators differ is that the cats get to eat whether they are successful hunters or not. Wild predators do not – they starve if they don't kill enough to eat, and so their numbers are kept in check by quarry numbers. Domestic cats lack this constraint and so are potentially far more damaging to wildlife populations than any wild predator ever could be. The same applies to feral cats kept alive by being fed by people, though feral cats that are entirely self-sustaining are subject to these economics of ecology and will only exist in numbers that the prey population can support.

While the numbers suggest that domestic cats can have an impact on wildlife populations, debate rages over whether cat predation does result in actual population declines in their prey species. Small birds and mammals have a high reproductive rate to allow for high predation rates. Each pair of robins may produce three broods of five chicks in a season, but if only one of the adults and one of the 15 young survives to breed in the following year, that still translates as a stable population. So perhaps cat predation rates are sustainable, removing only those weaker individuals that were destined not to survive anyway. Besides, domestic cats are mainly active in the urban environment, where many other predators are absent anyway. The RSPB's position is that cats do not have a significant impact on native wildlife populations, but the broadcaster Chris Packham is one of a number of prominent figures who have argued that real damage is being done, and that cat owners have a duty to do what they can to reduce the impact their pets have on wildlife. The usual counter-argument that 'it's just nature' doesn't hold up. While it is indeed in cats' natures to hunt – and perhaps, without hunting, their lives are lacking in some respect – their predation is not natural in ecology terms because domestic cats live outside of nature and so they are not subject to the normal dynamics of predator–prey relationships.

Keeping pet cats inside from dusk till dawn is one of the best ways of reducing their rate of predation on wildlife, if you can manage it, and this will also reduce their chances of being hit by cars. A belled collar is a more controversial option – the sound of the bell (or, better, two bells bumping together) as the cat moves can give quarry early warning and allow them to escape. But collars can snag and cause injuries or even death – even quick-release collars don't always work as intended. They usually do, but this means that an adventurous cat can lose dozens of quick-release collars over the course of a year.

The other, often ignored aspect of domestic cat predation on wildlife is the impact upon other predators. Every mouse or blue tit taken by a cat is one that's not available to a wild predator. This is probably of less importance in urban environments, where wild predators are already scarce. But rural pet cats – which are almost invariably more prolific killers than their city equivalents – could be outcompeting the likes of barn owls, weasels and kestrels.

What about feral cats? How and why do cats become feral in the first place? You only need to wander your local area for a while and you'll see that cats are somewhat prone to going missing, leaving distraught owners to stick signs to lamp posts and to leaflet-drop all the local houses. This happened to me in the 1990s, when my much-loved and favourite-ever cat, Jude, disappeared from our home in a Kent village. She was always inclined to wander, but after two days without sight of her we began to worry. I drew a picture of her (digital cameras were not a thing then) on a 'missing cat' note, and made a stack of photocopies of the note (home printers were not really a thing then either). We posted them through every letterbox in the village, and people started to call us over the days that followed, but nothing came of any of it. Although she was in fact half-pedigree – her mother a stunning lilac-point Birman with a flowing silky coat – Jude herself was a very ordinary, petite, short-haired black cat with white markings in the usual places black cats have white markings. There were plenty of cats just like her in the village and we must have been called out to inspect all of them.

I didn't want to give up on her. She had lived with us in four different homes by that time, and she had always been a roamer. This is unusual for female cats – they tend to be less adventurous than males – but that was her nature. Trying to keep her confined for any reason caused her to lose her temper and take it out on our other cat, Pixie, a docile creature who accepted these beatings as her due. Once I caught her hanging by her forepaws from the front-door letterbox, attempting to squeeze out. At a previous home, she would take herself off into the nearby woodlands for two- or three-day stretches. To bring her back we would walk into the woods at a time when no dog-walkers would be about, and Pixie would come too. We would walk the paths and call and whistle for Jude, and eventually we'd hear her sharp half-Birman yowl in the distance. She would come bounding out of the undergrowth to us, shaking dead leaves off her fur, and we would all head home – an eccentric two-human, two-cat party.

I suspect she fended for herself out in the woods – she certainly never seemed to be starving hungry when we brought her home. So I remained hopeful that she would be doing the same again, but time passed and soon I realised she'd been missing for three weeks. Hope didn't really seem like an option any more. I missed her, and regretted that I'd given her the freedom she had craved. Then there came one more tip-off from a local: someone had seen a small black cat hunting rabbits in the big field up behind the most distant of the village pubs.

We were there early the next morning, calling her name and whistling, just as we'd done to call her out of the woods. We skirted the treeline that bordered the field, flushing a little owl from its hidden perch in a big oak. It headed across our eyeline to alight on a fencepost, mocking us with its golden frown before departing in its comical bounding flight. No sign of any cat. We repeated our circuit, and then went home, despondent.

That evening, though, our quiet night in was interrupted by a yowl at the door. I was shaking as I opened the door. In walked my lost cat, staring up at me in her usual imperious way and demanding, in her usual insistent voice, to be lifted up and cuddled. As I picked her up, I could see that she had a few sheep ticks stuck to her face. Was that why she had come home? She certainly didn't feel as though she'd lost weight. She felt exactly the same as normal – stocky and strong, and purring robustly as she thrust her face and both front paws into my hair. I tried to imagine her battling and killing rabbits the same size as herself. It seemed so unlikely, but she had taken care of herself for three weeks. I was sure that if any person in the village had been feeding her, we would have heard about it. So rabbits it must have been.

This cat – ironically one of the most affectionate and human-friendly I've ever known – had what it took to go feral, even though she chose not to in the end. Pet tom cats that go unneutered are also very prone to wander away in search of calling queens, and they may end up fending for themselves many miles from home. But not that many feral cats are pets that have permanently strayed. Most were born wild and then grew up without human contact. They are the kittens of unneutered female cats who wandered, inevitably became pregnant, and chose to have their kittens outdoors away from people. If the kittens are never discovered, they will grow up as essentially wild animals even if their mother is someone's pet. In due course, these wild-born kittens will grow up and have their own kittens, and those will be feral too.

Jude and Pixie were the last cats I ever owned. Both of them were hunters, and I was finding it more difficult to square my love of wildlife with my love of domestic cats, as well as my belief that most domestic cats are happier if allowed outdoor access. The risks of letting cats roam were also brought home to me when we lost Jude on the road before her sixth birthday, only a year after her three-week rabbit-hunting adventure – and, though I was grief-stricken, I was unsurprised. There were suddenly many reasons not to have a cat of my own again.

Instead, I became involved with a local cat rescue that specialised in cats living on the streets and in other 'wild' situations. I offered short-term foster care for these cats in need. The second one I took on was the kitten Sookie, my *gatita fiera*, born feral to a stray mother and rescued just in time. Her story was one that came up again and again. After her, I fostered a mother cat who used to be a pet, and her born-feral kitten, who at five months old was full of charm but also full of fear. He was one of four and the most confident by far – the other three stayed at the rescue headquarters, too fearful to ever have a hope of a normal home. This cat family had lived around people and had been fed by them, so although my foster kitten was never handled, he at least recognised people as providers of food and not entirely threatening; over time he became much braver and almost tame. Next was an adult male cat who had lived wild for more than a year, but used to be someone's pet and retained a friendly nature. He had never left the suburb he called home, despite not being allowed into any houses and only sporadically being offered cat food.

The only feral cats and kittens that came my way for foster care were those that showed signs of being rehab-able. The wildest ones were beyond this but could sometimes be rehomed as farm cats. This involved trapping, neutering and vaccinating them, keeping them caged on the farm for a while until they had learned where their meals were coming from, and then giving them their freedom. With luck, they would stick around and reward their adopter by controlling rodents – but the deal had to include the farmer's agreement that cat food would still be supplied. Trap-neuter-return (TNR) is done at some feral cat colonies in Britain to keep the colony from growing larger and to improve the quality of life for the rest of the cats' feral lives. Many rescue charities around the world that run TNR programmes for feral cats will mark those that have been neutered by snipping the tip off one of the ears – ear-tipped cats are then released straight away if recaught.

Feral cats mostly live short, unhappy lives compared to their domestic cousins. A domestic cat is likely to make it to 14 years old and many do much better – most of us have known cats that have lived into their twenties. Feral cats are doing well if they live to the age of two, and reaching double figures is unlikely. They die from wound infections that would easily be treated by a vet, and from starvation, exposure, and road accidents. In some areas, they are shot or trapped as pests. Unneutered males fight one another; unneutered females are caught in a constant cycle of pregnancies that quickly exhausts their bodies. In North America, feral cats also have to contend with large, dangerous predators such as coyotes and pumas. Feeding a colony of feral cats can alleviate their suffering in the short term, but this also allows the colonies to grow and more kittens to be born.

TNR is a humane solution to a problematic feral cat colony, though some would argue that trapping and putting to sleep would actually be kinder than returning the cats to their difficult lives in some cases. Large baited traps are used to catch the cats, and once caught they are handled with extreme care as they present a real danger to the vets and charity workers who deal with them before and after their surgery. Cats this fierce have to be bundled into thick bags for safe handling, and given their sedative injection through the fabric of the bags.

Feral and domestic cats are potentially problematic in Britain, but in some other parts of the world their impact has been manifestly and unambiguously devastating. In Australia and New Zealand, cats have done untold damage to native wildlife. Cat curfews and other control measures are commonplace to help reduce predation on native wildlife, while feral cats are not tolerated at all and are killed in large numbers. When cats are introduced to small islands that are home to their own unique wildlife, the consequences are almost invariably bad. Island endemic wildlife species have often evolved in the absence of predators and so have no defences whatsoever against as accomplished a hunter as a cat. It's a widely told naturalist myth that the Stephens Island wren, a flightless songbird, was wiped out in its entirety by a lone cat over the space of a few weeks – the truth is that there was probably more than one cat, but the interval between this bird's discovery and its extinction was less than three years, and less than one year passed between the arrival of the first cat and the last known sighting of the wren. Worldwide, feral cats that have been introduced to islands have been responsible for at least 14 per cent of all of the world's documented bird, mammal and reptile extinctions over the last couple of thousand years, and today their presence seriously threatens almost 8 per cent of the world's Critically Endangered birds, mammals and reptiles.

Feral cats, it's pretty clear, are a problem for the whole world; so are domestic cats to some extent – the good that they do for us as companion animals notwithstanding. Their conflict with Scottish wildcats is just a small part of their story, but they are the key to why safeguarding our native wildcat is such a difficult undertaking.

The main problem is hybridisation. Some experts maintain that Scottish wildcats of both sexes would much prefer to mate with other wildcats, but feral cats and stray domestics vastly outnumber wildcats, and so a wandering tom or a calling queen wildcat is much more likely to find one of them than one of his or her own kind. Their behaviour is compatible enough that they will court and mate, and their genes are compatible enough that they can breed successfully. Their hybrid offspring can breed too – with wildcats, domestic cats and other hybrids. The scene was set long ago for a complete genetic mixing of the two, and that is what has happened. It's estimated by the IUCN that 88 per cent of free-living cats in Scotland are ferals, stray domestic cats or hybrids with a variable amount of pure wildcat genes. That leaves 12 per cent pure wildcats, and many wildcat conservationists would say that this figure is still much too high. The further it lives from human habitation, the more likely a Scottish wildcat is to be genetically pure, but feral cats can be tough and adventurous and have been shown to live and breed successfully many miles away from people.

This process isn't just happening in Scotland. Everywhere in Europe where European wildcats exist, there is the potential for hybridisation with feral and domestic cats to occur, and in truth hybridisation is probably extremely widespread. The IUCN's report on *Felis silvestris* goes on to say:

> *Wild Cats are most threatened by domestic cats. Although the lack of information, especially outside Europe, prevents us from drawing a general conclusion, hybridization is considered widespread; there may be very few Wild Cat populations remaining where there is little history of hybridization with the domestic cat.*

Hybridisation can happen in both directions – a tom wildcat would be attracted to a calling domestic or feral female, and a female wildcat could be mated by feral or domestic toms if no true wildcats are around to best them in a fight. The former is the more probable scenario, especially with young tom wildcats that are yet to find territories and so are more likely to venture closer to houses. Many Highlanders report their pet or farm cats producing litters of wildcat-like hybrid kittens that will probably end up living wild themselves if not handled and kept indoors. The hybrid offspring of a pure wildcat and pure domestic cat would be a generation-one (filial 1 or F1) hybrid. Its fertility would certainly be good enough to parent its own kittens, and if the other parent was also an F1 hybrid then their kittens would be F2 hybrids, and so on. A different naming system would apply if an F1 hybrid is mated back to a pure wildcat (or pure domestic cat), but with generation after generation of mixed-parentage cats freely interbreeding, any naming system quickly becomes meaningless. What we do know for certain is that in Scotland there are many free-roaming cats carrying a mixture of wildcat and domestic/feral cat genes. Some bear little or no resemblance to wildcats while others look almost indistinguishable.

The tricky business of telling a wildcat from a convincing-looking hybrid can be approached in a few different ways. One is by coat pattern. Dr Andrew Kitchener, the principal curator of vertebrates at the National Museum of Scotland, has been researching hybridisation between domestic cats and wildcats for 25 years, and has identified seven key traits that can be used to distinguish the two. They are as follows.

1 A wildcat has four stripes on the back of the neck.
2 It also has two shoulder stripes.
3 Its dorsal stripe (down the length of the back) ends at the base of its tail.
4 The tail has a broad black tip.
5 The tail is marked with neat rings (typically three to five of them).
6 There are seven to 11 unbroken stripes on the flanks.
7 The wildcat has no spots on its body sides.

The closer a wildcat comes to matching this standard, the higher its 'pelage score'. It's not the most accurate way of assessing wildcat 'purity' (that would be genetic analysis) but it's reasonably reliable and it can also be done quite non-invasively. All you need is camera-trap footage, though non-invasive ways of collecting material for DNA testing are also now in use.

While hybridisation is the main problem, feral cats can also harm wildcats in other ways. There are various infectious diseases that can affect domestic cats – some of them fatal. Some can be prevented by vaccination, and others may be treatable or manageable in pet cats, but obviously not so in feral cats or wildcats. There's not much evidence that this is a serious problem at present but wildcats can and have been infected with diseases such as feline leukaemia, feline coronavirus, which can develop into the lethal FIP (Feline Infectious Peritonitis), and FIV (Feline Immunodeficiency Virus). The cat owners in wildcat country who don't bother to vaccinate their pets or treat them when they are sick are exposing wildcats as well as their own pets to risk. Where trapping schemes are underway to catch and neuter feral cats, it's now the usual policy to vaccinate the cats too before release.

What about competition? This shouldn't be a problem. The old books tell us that Scottish wildcats are much bigger than domestic cats, as well as much stronger and fiercer. However, the pet and feral cats of today are not the same as those that were around when the old books were written. Nowadays we not only have cat breeds that are on average bigger than wildcats, but we have hybrid breeds like the Bengal (with a dash of Asian leopard cat genes) that can be very formidable and are notorious for starting – and winning – fights with other domestic cats. Most feral cats will not have these traits but some may. An exceptional feral cat with the right genetic background could potentially have the strength and aggression to displace a wildcat from its territory.

Domestic and feral cats add up to a big headache for wildcat conservationists. To solve the problem, the conservationists not only have to deal with the feral cats that are already out there and decide what to do about the many hybrids carrying a greater or lesser proportion of wildcat genes, but they also have to stem the flow of new cats going feral and moving into wild places. This means getting cat owners on side. Of course, many cat owners are completely on board with doing all they can to help safeguard Scottish wildcats, but cats also appeal to people who want a pet that can largely take care of itself. Finding a way to deal with straying and feral domestic cats – and intractable cat owners – is the challenge that conservationists today are facing, in their desperate bid to rescue the Scottish wildcat from its slow genetic erasure from the planet.

TRIP FIVE

Ardnamurchan, 2016

I'm glad to have a second chance to return to Ardnamurchan. This time it's spring – May, my favourite month – and a spell of exceptionally warm and fine weather is forecast. I'm not sure what to make of this, to be honest. The north-western Highlands of Scotland isn't really a sunseeker's destination. Will it feel strange to stroll those moors and lonely shores under blazing sunshine? Time will tell.

In Kilchoan, we get settled into the whitewashed bothy cottage that will be home for the next week, and then head down the road into the village. It's late, the light turning golden, and the shop is closed. We check out the bay – a wide beach, when the tide is out, made of gravel and boulders, cut through with little sea-bound streams and strewn here and there with heaps of washed-up, smelly kelp ribbons. The strandline attracts swarms of flies. The flies attract swarms of pied wagtails and rock pipits, and a common sandpiper trots across the big rocks close to the calm sea. Further out, common gulls are loafing on half-submerged rocks, and further still a few boats drift on the sparkling water. There is also a grey heron out there, stalking through the shallows on stilt legs, on the hunt.

A long, deep sleep later and we rise early to walk up to the hill summit. On the way we find a family of twites investigating the roadside – they fly up onto the fence at my approach and balance on the top wire, wobbling about as they eye me and consider whether or not to fly away. These are Highland birds, upland birds. Like many northern breeding birds, they leave the uplands in winter to head down the map and towards the coast, but they're still hard to see in my neck of the woods as only a few get far enough south. They are the plainest of finches in colour terms – just brown and streaky with a subtle blush-pink rump in the males, but their exceptionally sweet-looking faces are flushed a pleasing cinnamon tint and they have a sprightly manner.

Halfway up the road there are a couple of pigs rooting about in a muddy field, and on the other side of the road a buzzard is loitering on the remains of an old stone building, perhaps waiting to see if the pigs grub up any decent-sized earthworms. It takes off when we're too close and flies down towards the road, right over our heads, a dazzling sight with its broad, banded wings and tail, its dark eyes glinting. Up at the derelict house, a song thrush sings lustily from the rooftop, and the Atlantic gleams. Heading back down towards the village, we look up in response to a sudden uproar from the hooded crows, just in time to see a young white-tailed eagle sweep overhead, breathtakingly low and shockingly vast on its great sweep of wings, putting any and every buzzard to shame.

We head up the road out of the village and, after a good walk up into the moors, find an indistinct path leading into woodland. I'm thinking about wildcats now and walking softly, pausing frequently. What are Scottish wildcats doing at this time of year? Some females would have smallish kittens still in their dens, so it's not necessarily a great time to search, though those mothers will be needing to hunt much more than usual, to fuel their own bodies in order to feed their young. For the males – business as usual. We could bump into one in this little patch of forest.

We find our way out to a patch of high, heathery moorland. Here an old drystone wall is stumbling across the landscape, its base thickly lined with sedges and gorse. Beyond the wall, a small bird sits pertly on the highest point of a ridge of heather. It's outrageously pretty – a male whinchat, orange-chested with a boldly striped black-and-white face. He sings, his notes rather less pretty, and faint and lost in this big landscape. The dome of sky is immaculate blue, the air hangs still.

In the afternoon, we head out to Sanna Bay and walk and sit on the sand. I get busy with my camera, photographing ringed plovers and dunlins. The astounding weather has enticed a few humans to this remote and beautiful beach – no more than a dozen, dotted about on the shore and dunes, but it feels busy for Ardnamurchan. We take a walk beyond the bay, climbing increasingly steep and thickly vegetated hillocks of sand. A small, dusky butterfly flits at our feet. I track it until it settles. It snaps its wings shut and nearly vanishes – its blue-green underwings camouflage it perfectly. A green hairstreak, it's one of relatively few butterfly species that do well this far north.

Heading back, a couple of cars and a van have joined our own in the small car park, and standing beside the van is a man with waist-length black dreadlocks, a beard and a camera with a very long lens – the kind of cumbersome set-up that only a wildlife photographer would carry. He turns to us as we approach, ambles over, and we get talking. It's quickly evident that this man, Hamza, is highly knowledgeable about the peninsula and its wildlife and he's soon advising us where to go to see golden eagles, merlins, pine martens and more. When I mention wildcats, though, he goes quiet then turns on the image-viewing mode on his camera and begins to scroll through photos. I wait until he faces the viewscreen towards me. And there it is – a cat. A beautiful big brown tabby cat with a wide face and wide, angry eyes, with stripes and a banded tail with a rounded black tip. The cat is big in the frame, up against a fence of square mesh with young conifer trees behind it. I slowly raise my gaze to meet the man's smile. He describes where and when he saw the cat, how brief the sighting was. It looks as perfect a Scottish wildcat as I've ever seen. I feel like crying – but not with sorrow. My chances of a sighting like this remain as tiny as ever, but somehow knowing that someone else has had that moment, not very far from where I'm standing right now, is enough – is more than I'd hoped possible.

First thing the next day, we take the ferry across to Mull to meet friends for a full day of wildlife-seeking. No shearwaters to see from the boat this time but instead there are harbour porpoises and I've never seen them so well as I do today. These are miniature cetaceans, rarely more than 1.5m long, and they're introverts. No breaching or bow-riding for them. But the three that we see are lively enough, swimming alongside the ferry's steep flank and rising and falling, showing us their arched grey backs and, momentarily, their sweet, blunt-nosed faces.

Wildlife-watchers come to Mull mainly for the birds, and above all for the eagles. This island is perhaps the best place in the whole of the British Isles to see both golden and white-tailed eagles, and their supporting cast: hen harriers, short-eared owls, merlins, plenty of seabirds and plenty of landbirds of the kind that are more and more difficult to track down on the mainland. There are a few corncrakes here (though most people would hop one more island along and seek them on beautiful little Iona) and there are otters. Since at least 2012 and probably a few years prior to that, there are also pine martens. How they got here is not known – perhaps they stowed away on a boat, but they may have been introduced deliberately.

The presence of a small population of pine martens on Mull makes the island a bit more appealing for visiting wildlife-watchers, but there is concern that they could bring problems – they are, after all, very effective predators and are not native to this island. Scottish Natural Heritage is investigating the possible impact they could have on ground-nesting birds, and Mull Community Council would support localised control. Another non-native mustelid, the American mink, has already caused considerable harm to bird populations here and on other Hebridean islands and has been culled extensively.

When I've spoken to people involved with Scottish wildcat conservation, the idea of introducing wildcats to Mull has cropped up once or twice, simply because Mull is so strongly wildlife-focused. Tourists eager to see the eagles bring in an extra £5 million or more to the island each year and support more than 100 extra local jobs. The chance of seeing one of Britain's rarest and most elusive mammals would elevate those figures to even more impressive levels, but it's probably a no-go as there is no confirmed historical precedent for an established wildcat population on Mull, even though there have been occasional possible sightings. Reintroductions are one thing. Brand new introductions are another entirely.

We meet our friends at the Tobermory port and drive out of town. It's still early but the heat is building as we head towards the wilder, more rugged western side of the island. The improbable sunshine lights up a stunning male cuckoo perched on a wall at the roadside – we reverse carefully, and the cuckoo stays put just long enough for me to stick my lens out of the window and take a few photos. I've never seen a cuckoo at such close quarters before. His indignant expression, thanks to a staring golden eye, belies the otherwise hawkish impression he gives. He's soon off, a gangling long-winged, long-tailed shape flying haphazardly across the field beyond. Male cuckoos, as revealed by satellite-tracking, may spend as little as two months in Britain before beginning their return migration to Africa, and they have a lot to get done in that brief window of time. They spend only 15 per cent of their time in Britain, compared to 47 per cent in Africa; the rest is spent on the road in the two annual migratory journeys. These follow a round-trip path that loops down to Africa via Italy but returns on a more westwards track through Spain.

We visit a bay and reedy river, and listen to the fishing-reel song of a grasshopper warbler from somewhere deep within the waterside vegetation. Common gulls float like angels over the bay, and waders explore rocks and muddy ridges. I feel I could stay here all day, but to see Mull's eagles we need to be elsewhere.

White-tailed eagles were originally introduced to Rùm island but have gradually spread through the Inner and Outer Hebrides as well as to parts of the mainland. They have done particularly well on Mull, with 20 pairs here in 2017. Golden eagles are present too, in similar numbers, though they are more difficult to see as they favour the higher, more remote inner parts of the island while the white-tails – eager fish-eaters – stick closer to the coast. We park up by a sea loch and wait a while, scanning sky and shore. There are other birders here doing the same. No eagles appear, and after a while I wander off to photograph hooded crows and mistle thrushes in a steep field opposite the road, but am then alerted by a shout from my companions to look up.

The eagle soaring overhead is high up, diminishing its magnificence a little, but through binoculars I can see it in full, detailed glory. Despite its great size it is gracefully proportioned, its outline balanced. That's how I know at once that I'm looking at a 'goldie' and not a bigger but somehow ungainly white-tailed. It is a subadult, with patchy bands of white across its tail-base and the midline of each wing. It's also developing the golden nape of adulthood, and its huge yellow feet bunched up against the tail are prominent even at this range. It describes a series of widening circles as it gains height, before drifting across the sea loch and away. Hot on its heels comes a magnificent adult white-tailed. Its wings are not elegantly angled like the goldie's but are great oblong slabs fringed with feather fingertips. The little wedge-tail is pure white, the head almost white too and contrasting with the huge yellow bill.

Our two eagles are not close relatives – they belong to different genera and have different ways of life. To me, the white-tailed eagle is like the wolf, while the goldie is like the wildcat. Wolves and white-tails are sociable – the wolf obviously much more so – but it's not unusual after fledging for young nest-mate white-tails to associate for months, foraging and playing together as they wander in search of a permanent territory for themselves. White-tails will also converge at a rich food resource, though they may bicker as they do so. Sometimes goldies might turn up too but they are usually displaced by the bigger and burlier white-tails, though a mid-air scrap might suit the more agile goldie better. Golden eagles are not social with their own kind in the least. They don't even usually have nest-mates, thanks to the habit of cainism, a trait of the *Aquila* eagles whereby the first-born chick repeatedly attacks and eventually kills its younger sibling.

White-tails are also fairly relaxed around people, which admittedly wild wolves are not. Some of the wildlife tour companies organise boat trips where guests can photograph white-tailed eagles at close quarters as they swoop down to scoop up thrown chunks of fish, but you would never catch a golden eagle carrying on in this manner. Like wildcats, they prefer to keep themselves to themselves – they nest far away from habitation and hunt and forage in the wildest places. Sadly this isn't enough to keep them safe on the mainland – they are much persecuted on grouse moors and are missing from many regions that hold ample habitat to support them. In my mind, the golden eagle, capercaillie and wildcat make up a trinity or triquetra, some top tier of wildlife magic that's representative of the most mythical wildernesses of Scotland. They hang on in their determined isolation as humanity presses in on their last refuges. Even if you manage to glimpse one, the sense that it's still out of your reach is undimmed. But now, they are fighting for life, all of them.

Our day on Mull is over too soon. We take a late ferry back, and then spend another evening in a slow and fruitless search for wildcats out on the hills. By now I'm used to the half-seen ghosts that stalk through the shadows, the wildcats that aren't there. The deer herds shift slowly out from the woodlands into open ground as the light fades, and we drive back in midnight twilight, through the sleeping village.

Glenborrodale is an RSPB reserve on the south shore of the peninsula, and I see that its RSPB webpage mentions wildcats, albeit in non-committal tones. We're there early, but is it early enough? The sun's there before us once again, and it's already hard going in the heat as we walk a climbing trail through the hanging oak woodland. The trees are extravagantly twisty and lavishly strung with lichens. The surrounding terrain is covered in boulders but many of their outlines are softened by a shroud of thick moss; in between them heather and bracken grows thick and tufty. It's easy to imagine a wildcat negotiating this landscape. A distinctive song

rings from the high trees – sweet and shivery. I know
this singer and track it down among the sunlit fresh
canopy leaves after a little patient scanning. A wood
warbler, the biggest and brightest and scarcest of our
breeding *Phylloscopus* or 'leaf' warblers. He is stunning,
decked out in leaf green and lemon-yellow, with a
silver-white tummy and an elegant structure. His long
pointed wings, which reach nearly to his tail-tip, vibrate
as he sings.

The woodland gives way to open moor, and we can
see down to Loch Sunart in all its calm beauty. A
butterfly pauses to warm up on a bare rock. It's dark
grey-brown with big creamy spots – a speckled wood.
I know this butterfly well from back home, but the ones
here are different, their colours more extreme. A Kentish
speckled wood looks brown and yellow, but this one is
much more contrasty – black and white almost – and its
spots are bigger. It's of a different subspecies – *oblita* –
which lives here in north-west Scotland. The ones back
home are the subspecies *tircis*. In the last couple of
decades, the speckled wood has expanded its range,
taking advantage of countryside changes that have
created more of the grassy woodland-edge habitat it
requires. It's increased in abundance by an amazing 84
per cent since the mid-1970s and spread its range by 71
per cent over the same period, bucking the general trend
for butterflies. There are now populations of *tircis* in
northern Scotland. Some are almost in contact with
oblita. It's pretty clear, looking at the map, that *tircis* and
oblita will soon be intermingling quite extensively, but
what will that mean for them both?

I'm pondering this, and wondering if parallels can be drawn between this butterfly and the Scottish wildcat's travails with domestic cat hybridisation. It sounds absurd at first glance, but it's possible that *tircis* could dilute the genetic uniqueness of *oblita* to the point where the latter would no longer exist in any meaningful way, at least on the mainland (whether *tircis* could reach and colonise the various western Scottish islands where *oblita* occurs is another matter). What's different is that these two are unambiguously different subspecies of the same species, whereas the Scottish wildcat and the domestic cat belong to two different (albeit very similar) species, according to most taxonomists. The other important difference is that the *tircis* speckled woods have expanded and spread their range under their own steam rather than through direct human intervention.

But I'm not sure how important these factors really are. Whether we call a distinctive form of an animal a species or a subspecies doesn't mean anything much in nature. Arguably, the precise reasons behind a species' population expansion don't mean much either. We may not have physically brought *tircis* to the Highlands, but it was our activities that set up the environmental changes that allowed it to happen. A similar scenario is unfolding in North America, where deforestation in central states is allowing the barred owl, which is native to eastern parts and does well in more open habitats, to spread into the range of the spotted owl, which is western and likes mature, more closed woodland. Where these two closely related species meet, there are problems – mainly through the barred owl outcompeting the spotted for nest sites and territory, but hybridisation is happening as well. There are no moves here in Scotland to protect *oblita* from the encroachment of *tircis*, but spotted owls are subject to a suite of conservation measures in North America. Is there really a difference?

Once a wildlife-related conversation moves around to matters of genetic purity, other thoughts inevitably come to mind and they're not comfortable ones. The history of our own species is littered with shameful examples of regimes that sought to segregate or eliminate those with a different genetic make-up to the preferred 'type'. Can any parallel be drawn between this and our efforts to protect certain populations of wild animals from genetic dilution?

It's an interesting philosophical question to chew on mentally as we carry on with our walk. I feel strongly that saving a unique animal is a worthwhile enterprise under virtually all circumstances, but particularly when it is our carelessness that has put it in danger. At the same time, I've loved cats all my life and instinctively baulk at any plan that would harm them, be they domestic or feral. In the case of the Scottish wildcat, though, the recommended actions for domestic and feral cats to further wildcat conservation are measures like neutering and vaccination, which will also benefit the domestics and the ferals themselves. It would be a difficult dilemma, though, if the only solution to the Scottish wildcat's problems were a wholesale cull of feral cats and a curfew (with deadly consequences if ignored) for pets. I recall talking to an RSPB seabird biologist about his work on remote islands, where invasive predators like feral cats could destroy entire colonies of petrels and other birds in a matter of weeks. He told me that he liked cats, but if he found one on one of his islands he would destroy it without hesitation. This kind of clear-eyed pragmatism is essential sometimes for real-world conservation, but not many people have it in them to manage it.

I am distracted by a latticed heath moth flying over the sunny heather. In flight it looks intriguing with its spotty brown-and-gold wings but it's only when it settles that I see what it is. I'm disappointed because there is a butterfly living here that looks a bit like that and is one I'd really love to see. The chequered skipper became extinct in England some decades ago, and now survives only in this area of north-west Scotland. We're in the middle of its flight season. It is a mercurial and finicky character, only showing in the right weather conditions. Today actually couldn't be more perfect for butterfly activity, so I don't know why it's not out and about, but it's not.

The chequered skipper is another species which will hopefully be restored to its old haunts very soon through reintroduction. The year 2018 marks the first release phase of the project. The butterflies in question were caught in Belgium and rushed over to a secret area within Rockingham Forest, Northamptonshire – a site that has already seen a successful red kite reintroduction. The habitat at the site has been assessed and improved to suit the butterflies, and of the 42 insects released, 31 were females that would already have mated and be ready to lay eggs. If all goes well, a new generation will be seeded, and the spring of 2019 will hopefully see the first home-grown chequered skippers to fly over English soil since 1976.

We're close to the end of our walk when I pause to look at another moth, and as I'm following it as it flits about in a dappled clearing, I suddenly realise I'm staring straight at a tawny owl that is perched in a small tree just a few metres away, exactly at my eye level and staring back at me with its unreadable gleaming black gaze. I am so surprised that I would have dropped my camera if it wasn't on its strap around my neck. The owl sits motionless, perhaps wondering if its exquisite camouflaged pattern of russet, ash and cream spots, streaks and bars is doing its job. When I lift my camera and take a couple of photos, it concludes that no, I am looking directly at it and it steps off its branch to fly away from me. It doesn't go far, though, alighting on another fairly nearby tree and turning to stare a little more. It makes me think of dryads or other woodland spirits – I feel I shouldn't have seen it, that it should have stayed hidden from my gaze through some timeless magic. Luck of this magnitude is what I'd need if I were ever to see a wildcat. Today, I'm grateful and enchanted by this owl. It's not a rare bird at all but a sighting like this – by day, at such close range, in high summer – is extraordinarily rare.

We're drinking in the bar in Kilchoan's only hotel that evening when Hamza strolls in. He agrees to join us for a drink, and we talk. An African native, he is permanently settled here in Ardnamurchan now, and has a fund of tales from years of wildlife-watching and photography. However, for all his long hours in the field, he has had only the one wildcat encounter. And, much as talking to him makes me wish I could stay here forever, I feel more and more clearly that forever might not be long enough to meet a Highland tiger. I've always felt like a lucky person, but the level of fortune required for this particular experience is not comparable to anything else. Maybe in 20 years' time the various conservation efforts underway will have paid off and there will be wildcats here in abundance. Well, probably not abundance – that's never likely to happen – but in numbers good enough that a tourist here for a week would stand half a chance of a sighting. I hope so, even if that tourist is not and never will be me.

Our time in Ardnamurchan is rounded off with a boat trip out to some of the more remote islands hereabouts. We spend a morning on Lunga in the Treshnish Isles, communing with puffins on the high grassy tops. Then we cross open sea, on the way meeting a vibrantly lively party of bottle-nosed dolphins that bow-ride with us at point-blank range for ages and send the boatload of humans into dizzy delight. Finally the dolphins leave us and we approach Staffa, slowing down to a chug to admire its shores with their remarkable basalt pillars – each a perfect hexagon in cross-section – to peer into the narrow, deep and booming cavern of Fingal's Cave, which inspired one of Mendelssohn's best-known pieces, and finally to land on the island itself. We seem very far from the mainland here, and very far from wildcat potential as we explore the treeless green crown of Staffa. But there is a surprise rarity hiding here: from what seems to be the island's only patch of decently long grass comes the sore-throated, rasping call of a corncrake, a bird nearly as rare as the wildcat itself and nearly as elusive except for that telltale song.

Scotland has yielded so many of its gems to us this week. Once again, not the big one, but time here in this beautiful hidden corner of the Highlands is never wasted. As I listen to the corncrake's voice, I remember Hamza and his wildcat photograph, and wonder whether he'll ever repeat the experience. I hope that he does.

CHAPTER SEVEN

Salvage Operation

We're not good at recognising our mistakes as a species. Nor are we terribly good at making amends, even if it's not already too late (which it often is). The wildcat of Britain underwent centuries of systematic persecution with almost no dissenting voices. In Mesolithic times its population in Britain was an estimated 66,033 individuals, according to 'The Mesolithic Mammal Fauna of Great Britain' by Maroo and Yalden (2000). Today, population estimates of pure Scottish wildcats still surviving range from a generous 2,000 to no more than 35. In any case, the decline has been catastrophic and the contraction in range perhaps even more disastrous. And remember that it was not until 1988 – when all this damage was well and truly done – that the wildcat was finally awarded full legal protection.

I nearly used the word 'exceptional' to describe how the wildcat's numbers plummeted, but sadly this is *not* an exceptional situation. Our history of sharing our land with predatory mammals and birds is shameful. Barely any species has escaped relentless destruction – not even species as innocuous as the kestrel or the short-eared owl. Only since the mid–twentieth century has it become a mainstream view that it is worth our while protecting and encouraging our populations of these species, and that predators have an actual right to exist. This change of view can in no way be relied upon to prevail in the long term. Today, with conservation efforts bearing fruit in some cases, the bad old ways are rearing their heads once again, with calls from various quarters to cull otters and pine martens, red kites, sparrowhawks and peregrine falcons. The entrenched, visceral antipathy that some people hold towards wild predators has not gone away, and it probably never will as long as these wild animals threaten our own interests. Even those of us who don't want to fish, shoot or keep pigeons and poultry, and those of us who pride ourselves on cherishing nature unreservedly, probably aren't immune. If wild predators began to affect our chances of feeding our families, who's to say we too would not turn against them?

Another question around wildcat conservation is this: do we really *need* wildcats? In matters of ecology, can't we just leave things as they are, let nature take its course, and allow the feral and hybrid cats to occupy their niche? In a very pragmatic way, yes, we could. But feral cats differ from Scottish wildcats in several important ecological ways – most significantly in their tendency to aggregate into colonies wherever they have access to plentiful food. Nevertheless, there is more to this than ecology and pragmatism. This is also about saving something unique and irreplaceable, and about undoing the wrongs of our less enlightened past. As a nation we are belatedly learning to value our wildlife – even that which can cause us problems – and we're taking responsibility for putting things right.

The optimist in me thinks (or hopes) that this sea change in attitude can be sustained, at least for a majority of us Brits. We have managed to hang on to the ban on hunting with dogs – for now, at any rate, although it *is* constantly violated. Raptors are still illegally killed on grouse moors and pheasant-shooting estates but the voice of the outcry grows louder each year, with Parliament now beginning to pay attention. The government has so far resisted anglers' calls for an otter cull, and a range of raptor reintroduction projects have proceeded – and succeeded – despite public opposition in some areas. There's no denying that we are starting from a low point. The UK and other countries were recently scored for biodiversity 'intactness', assessing how badly harmed wildlife populations have been by human activity. The UK was ranked 189th out of 218 countries, a dismal record indeed. Yet we are, in theory, in a strong position to put things right or at least to make things a bit better.

Restoring populations of rare wild animals can be effected in various different ways. Reintroduction is perhaps the most obvious, though far from the most simple. In Britain, attempts have been made to restore populations of red kites, ospreys, white-tailed eagles and pine martens, as well as various other non-predatory species. The red kite scheme in particular has been a roaring success, although those in England have fared far better than those reintroduced in the eastern Scottish Highlands, where illegal raptor persecution is still commonplace. It's early days but the white-tailed eagle reintroduction to eastern Scotland is also not proceeding as smoothly as the earlier west-coast attempts, with persecution a factor.

To make a reintroduction work, you need to do a lot of groundwork. Simply letting animals out into the wild is no good. In the 1970s and 1980s, many captive-bred barn owls were released in England in a vaguely coordinated but misguided attempt to prop up their falling population. The released birds were mostly flying to their doom because the factors causing the decline of the wild owls had not been addressed and because the released birds were not being prepared properly for a life in the wild. When barn owl numbers began to recover around the turn of the century, it happened not because of reintroductions but because life became easier for the owls already here. Agricultural changes started making it easier for them to find food, and people began to put up owl nest boxes, which made it easier for them to breed successfully (many former nest sites were lost in the rush to 'do up' old rural buildings through the late twentieth century). No reintroduction is likely to work unless the project coordinators have both identified the causes of the original decline *and* either dealt with them or established that they are no longer in effect.

The red kite reintroduction is a great example of a well-planned project of this kind. Kites were wiped out in England and Scotland through persecution and, latterly, egg collection. Thanks to legal protection and attitude changes (including a fuller understanding that these birds are mostly scavengers and no threat to human interests), it was reasonable to suppose that serious persecution would no longer occur. And the habitat that the kites would need was still plentiful – these adaptable birds do well in the sort of farmland/woodland mosaic that covers much of lowland Britain.

The next piece of the puzzle was finding some actual kites that would be suitable for release here. There were still red kites in Britain – a few pairs had managed to hang on in Wales – but for this project it was decided to use birds from Spain, which has red kites in abundance, instead of depleting the small Welsh population. For the concurrent reintroduction in north Scotland, though, birds were taken from Sweden. This difference reflects the fact that kites in Wales don't migrate and so English ones probably wouldn't either, whereas kites in Scotland might because kites from equivalent latitudes in mainland Europe do. In birds, migratory behaviour (or the lack of it) is frequently hard-wired into the genes, so it made sense to use stock with a propensity to migrate south for the Scotland project, and stock from a sedentary population for England.

Accordingly, as the project began in earnest in 1989, conservationists took kite chicks from nests in Spain and Sweden, and reared them close to the release sites in aviaries, keeping human contact to a minimum. When the young kites were flying and feeding themselves, they were allowed their liberty, though food was still provided for them until it was no longer needed. They were also marked with distinctive coloured and numbered wing-tags, which could easily be seen in the field and allowed the conservationists to track the kites' movements. More young kites were released in the same way over the next few years – nearly 200 of them at each site in total. The outcome today is that there are more than 1,000 pairs of red kites breeding in the area around the Chilterns release site, while the north Scotland site is getting on for 100 pairs. Overall, red kites are doing well, with their UK population further boosted by several more reintroductions at other suitable sites. And it also transpired that the Scottish kites did not show migratory behaviour, despite their Swedish heritage.

The successful reintroductions of ospreys to England, pine martens in Wales, and white-tailed eagles in Scotland and Ireland have followed broadly similar protocols. There is also much talk of reintroducing the Eurasian lynx to Britain. In July 2017, the Lynx UK Trust submitted a licence application to release lynxes into the UK, and in 2018, landowners in Kielder Forest agreed to play host to the first reintroduced animals. This latest scheme is by far the most ambitious, but is well supported and backed by robust science. Although this initial application was rejected by the government in December 2018, the Lynx UK Trust remain optimistic that a revised proposal will be successful. Over the next few years, we are likely to see the return of a truly iconic wild animal with a vital ecological role to play – even if, when it comes to public perception, this would surely be the riskiest reintroduction yet.

Reintroduction of Scottish wildcats will happen one day, but not yet. The threat they face from domestic cats is still very much present over most of their natural range, so the first hurdle remains in place, for now at least. But Scottish wildcats are in a much more desperate situation than ospreys, pine martens, white-tailed eagles and red kites. Those species are all well represented elsewhere in Eurasia (indeed, worldwide in the case of the osprey); however, for Scottish wildcats, the population in Scotland is all that we have (except for those residing in zoos and other captive arrangements). It's also important to note that, in mainland Europe, reintroduction of European wildcats has been attempted in some countries but never with any great success. For Scottish wildcats, reintroduction isn't currently an option but it will be some day, and in the meantime there are other conservation efforts underway.

In most cases, the best thing you can do to protect a species is to look after its remaining wild population. If you can make sure that its habitat is kept intact and then perhaps expanded, and that threats are kept away, it might be enough. It proved enough in the case of the marsh harrier, a beautiful bird of prey that hit a low of just three birds (a male and his two girlfriends) in the 1970s but today is on 400 pairs and still increasing. Legal protection from disturbance and persecution – coupled with preservation and expansion of reed beds along the eastern and southern coasts of England – allowed this impressive recovery, and no reintroductions were needed (though it's likely that the growing population was also boosted by a few opportunistic visitors from mainland Europe).

For Scottish wildcats, habitat destruction isn't the main problem right now – there is ample suitable habitat still remaining across the Highlands to support a good wildcat population, and ongoing reforestation projects should also help wildcats. Of course, any areas that can be shown to hold wildcats should be particularly carefully protected from damage or disturbance. Identifying such key areas is not easy, though. Surveying for such an elusive animal takes time, patience and luck. Camera-trapping is the best way. Putting out bait in the right place should encourage any passing cat to give the camera a good, clear view, enough to see all seven of those crucial bits of its pelage that help separate hybrids from the real thing. Each of the seven traits is scored out of three for how closely it comes to the pure wildcat 'standard', giving a total potential pelage score of 21. Cats that score 18 or higher are deemed to be high-scoring and close to pure – it's a high bar but for a good reason, as preserving the true essence of the Scottish wildcat is the goal.

Ideally the bait is placed above ground level so that the cat can be photographed standing up on its hind feet or jumping, both of which should provide a good view of its tail. Researchers are also working on ways to trap some of the cat's hair, so that DNA testing can be carried out. Norman Davidson, a volunteer for Scottish Wildcat Action, is (at the time of writing) developing a tunnel device with a movement-triggered coil that pinches a few hairs from the passing animal, and which so far looks very promising.

Camera traps take some practice to use correctly. Without careful siting, they can be triggered repeatedly by wind-blown vegetation. They will also, inevitably, photograph other wildlife, which might be interesting (who wouldn't like to see a pine marten or a capercaillie on their camera?) but also frustrating. Trail cameras are also sometimes stolen or vandalised by people. Of course, a more likely outcome is that they capture nothing at all, which is far from unusual in the challenging world of Scottish wildcat conservation. But camera-trapping has also had great success, and has led to the identification of several areas where wildcats (or wildcat-like hybrids) are still present and where the habitat is extensive enough to support them in sustainable numbers. These have been designated conservation priority areas – they are Morvern, Strathpeffer, Strathbogie, Northern Strathspey and the Angus Glens, and current wildcat conservation is very much focused on them.

Deliberate persecution is not a serious threat at present. Wildcats have full legal protection – they cannot be killed or harmed, nor disturbed at their dens. It is possible that some wildcats are mistakenly killed by gamekeepers and landowners targeting feral cats. This difficult problem is best tackled through education. When you are out lamping at night and you pick up a cat in your beam, you're unlikely to be in any position to make a meaningful decision about whether it could be an actual wildcat before the opportunity to shoot it has passed. Using snares is even less discriminatory. Therefore, while lamping and snaring are both legal, a better option is to use a humane cage trap. This way, a caught cat can be inspected properly without any harm. Scottish Wildcat Action, one of two key projects involved in wildcat conservation in Scotland, advises anyone active in pest control in wildcat country to use this method.

The most crucial aspect of Scottish wildcat conservation right now is dealing with the issue of feral and (to a lesser extent) domestic cats. This intractable problem requires a multipronged approach. First of all, the flow of new cats into the feral population needs to be curtailed, which means spreading the word that cat owners should neuter their pets. For many – but not all – pet owners, this one is a no-brainer. One of the conservation biologists I spoke to told me of a woman in her study area who owned numerous free-ranging, unneutered cats. She defended her cats' 'right to roam' and freedom to procreate at will. She also harboured an intense dislike for Scottish wildcats, and didn't care about their desperate predicament at all. When up against this level of resistance, conservationists have to invoke the many benefits to neutering for the pet cats themselves: greatly improved health, and greatly reduced chances of straying and fighting. These arguments finally did the trick in the case of this particular recalcitrant owner.

Then there is the need to deal with the feral cats that are already out there. In legal terms, you are still allowed to kill feral and stray cats on your land, as well as wildcat–domestic cat hybrids if you want to – and many landowners do. But both of the major wildcat conservation projects underway in Scotland today go with the trap-neuter-vaccinate-release (TNR) method instead. A neutered, vaccinated feral cat poses no risk to Scottish wildcats, and its territoriality may also help to keep away new feral cats from an area.

The other major component to wildcat conservation lies in captive breeding. This is a vital tool for conservationists dealing with a species in dire straits, although the costs and effort involved are considerable. When a species is declining fast and the causes can't be immediately addressed, maintaining a population in captivity is a safeguard against extinction in the wild. It can also be a source of stock for future reintroductions. Captive breeding wasn't necessary for the red kite and white-tailed eagle reintroductions because both species had strong populations in other countries and so young birds could be taken from the wild there. But that's not an option for animals with a tiny remaining global population – such as the Scottish wildcat.

There is a long list of animal species that have been saved through captive-breeding schemes. Some famous examples include the California condor and the golden lion tamarin (a striking fiery-furred monkey from South America). Some other animals that are extinct in the wild today still exist in captivity and are being bred to build a stock with the hope of eventual reintroduction to the wild. These include Spix's macaw, a stunning blue-plumaged parrot that inspired the children's movie *Rio*. Captive breeding not only keeps the species safe, but conservationists can keep full control of the breeding pairs so that genetic diversity is maintained as much as possible, and can give the offspring a much better chance of survival than they would naturally have when growing up in the wild. Captive breeding and release is turning around the fortunes of the Iberian lynx, one of the world's most threatened cat species, and a successful programme of captive breeding has also helped shift the Amur leopard, a threatened subspecies, just a few steps away from extinction.

Yet this isn't an option for everything. Not every wild habitat can be recreated adequately enough to encourage animals to breed. The alternative in such cases might be to take young animals into captivity and hand-rear them but this isn't necessarily straightforward either. I recently researched the conservation efforts underway to save the hooded grebe (a beautiful Argentinian wetland bird) from extinction. This bird breeds in groups on wind-stirred lakes on high plateaux. Captive breeding is a non-starter, but so far the attempts to hand-rear young have also failed.

Happily, Scottish wildcats can be bred reasonably successfully in captivity. They are much less fecund than their domestic cousins, but a compatible pair should be able to produce a litter each year, and to continue to breed until they are nearing double figures – provided they remain in good health. Introduce the male to the female while she is on heat and nature will probably take its course with only minor snarling and fisticuffs.

The Scottish wildcat captive-breeding programme is a coordinated effort between several zoos and several other facilities that are not open to the public. Every wildcat that is considered for the programme has been scored for both its pelage (see above) and its genetic make-up, and only those that score highly on both fronts will be considered for breeding. The genetic test developed by researchers for Scottish Wildcat Action takes as its starting point a pure wildcat genome that has been mapped by using DNA taken from old museum specimens as well as a rather more easily obtained domestic cat genome. When a putative new member of the breeding programme is tested, the researchers compare its genome to that of the wildcat 'standard'.

To get a little more scientific, the test singles out particular 'SNPs' for comparison. An SNP is a single-nucleotide polymorphism. Nucleotides are the molecules that make up a strand of DNA and there are four types in mammal DNA: adenine (A), cytosine (C), guanine (G) and thymine (T). In a DNA double-helix strand, they come in pairs, and A is always paired with C, and G with T. Comparing DNA strands of two individuals of the same species, you'll find the same types of nucleotides in the same places as you look along each strand of DNA that makes up each chromosome. In the case of an SNP, though, there's variation – some individuals have an A at that particular site on the chromosome and others a C. These variations in DNA give rise to variations in the animal's actual body because DNA is the set of instructions that tells an animal's body how to grow. The wildcat genetic test devised by Scottish Wildcat Action looks at 35 known SNPs in nuclear DNA. (This explanation is a colossal oversimplification, but if you are minded to do so you can get properly stuck into the science by reading the paper 'Wildcat Hybrid Scoring For Conservation Breeding under the Scottish Wildcat Conservation Action Plan' by Dr Helen Senn and Dr Rob Ogden, which is available online.)

Applying the genetic test generates a 'Q score' – for a hypothetical, entirely pure wildcat the score would be 1, and for a domestic cat it would be 0. So the closer to 1 any cat scores, the more wildcat genes it carries. In general, only cats with a Q score of at least 0.75 are considered for breeding, and much higher is preferred, but selection of breeding stock takes the cats' pelage scores into consideration as well, though genetic scores are more important overall. To give an example, the two breeding pairs of wildcats that are currently kept at Wildwood in Kent are scored as follows:

Pair 1:	Carna (F)	RJ (M)
Q scores:	0.983	0.984
Pelage scores:	16.20	19.20
Pair 2:	Isla (F)	Jura (M)
Q scores:	0.979	0.803
Pelage scores:	14.15	17.19

These figures show that a high Q score doesn't necessarily correlate with a high pelage score. They also show how close some captive wildcats really are to being genetically pure. Carna and RJ are very high-scoring, and have also proved compatible, with a trio of kittens born to them in 2017. The quirks of inheritance can mean that kittens could show lower (as well as higher) Q scores than their parents, but as both Carna and RJ's scores are so high, their kittens should all fall easily above the threshold and enter the breeding programme in due course.

Once you have determined whether a cat is wildcat enough to breed from, you need to find a suitable partner for it, and here the most important thing is selecting pairs that are not too closely related. Inbreeding depression – where genetic flaws are concentrated and magnified in the offspring of closely related parents – is a real hazard when dealing with a small population of animals. It's therefore crucial that detailed records are kept of each cat's ancestry. Keeping track of this is known as running the 'studbook', although I wouldn't like to attempt it today using an actual book; this is surely a job for a complicated and multicoloured spreadsheet. There is full cooperation between the various breeding facilities, and the cats are moved to where they need to be to meet new partners in good time for the start of each breeding season.

Talking to staff at Wildwood reassures me that the keepers are in no danger from their wildcat charges. I already described how Isla, a hand-reared female, greeted her keeper with the sort of purring ecstasy you'd expect from a friendly moggie, but even the wilder wildcats have never been known to attack a keeper, and there have been wildcats at Wildwood since 2002. They will hiss and growl and keep their distance, but only when the keepers need to actually handle the cats are they at risk of getting hurt. You might think (if you didn't think too hard) that habituating the cats a bit more might make their lives in captivity less stressful, but their essential wild nature needs to be preserved, for the sake of future generations that will hopefully be returned to the wild. Far better to keep things as natural as possible.

The Scottish wildcat is not a tremendous draw for visitors to the zoos. It is for *me* but often when I've spent time watching zoo wildcats I've been there alone. They just don't seem to draw the crowds, and the Wildwood staff have confirmed my hunch about this. Their enclosures tend to be sited a little off the 'main drag', which is partly to keep things quieter for the wildcats but also because people just aren't that interested. This is another element of the fight to save this cat – conservation efforts are much more likely to succeed if they have the backing of ordinary people. So why don't ordinary people care that much about wildcats, and how can we change this?

The main problem is one of appearances: there is no getting away from the fact that wildcats *do* look a lot like pet tabbies. It's easy enough to impress people with photos or artworks but wildcats in zoos are not going to spend their days striking dramatic poses for the cameras, and crouching at bay with ears flattened and fangs bared – and nor should they! Maybe it is just as well that zoo wildcats aren't too popular – they are more likely to be comfortable with fewer visitors, after all. But that doesn't mean people shouldn't know their story.

When I visited the Highland Wildlife Park's wildcats, Hamish and Betty, I happened to be there just as a young keeper was giving the cats their food, and she talked to the visitors about the cats at some length. It was clear that several of the people listening to her had not known of the existence of the Scottish wildcat before – and this was in Scotland, in actual wildcat country, where human actions and inactions can make a real, tangible difference to the creature's ongoing survival. For people living at the opposite end of the country, knowing about Britain's last native wild felid would be even further off the radar.

However, news stories about wildcats do tend to make it onto the main UK news pages, whether they concern new kitten births in zoos, particularly fine wildcats caught on camera traps, or the latest developments with conservation and reintroduction plans. Following these up online will lead you to the websites of the two main conservation projects: Scottish Wildcat Action, and Wildcat Haven.

Scottish Wildcat Action, affiliated with the Royal Zoological Society of Scotland, is a collective of some 20 organisations that coordinates wildcat captive breeding, as described above. It is also involved in *in situ* conservation efforts including surveying for wildcats using camera-trapping, and making a concerted effort to trap, neuter, vaccinate and release feral cats and low-scoring wildcat hybrids. If very high-scoring cats are found, these may be caught and added to captive-breeding programmes in order to increase genetic diversity. The organisation also campaigns for wider neutering and vaccination of domestic cats. It is publicly funded. Besides the Royal Zoological Society of Scotland (RZSS), the partner organisations include Scottish Natural Heritage, Cairngorms National Park Authority, Forestry Commission Scotland, National Museums Scotland, National Trust for Scotland, Royal (Dick) School of Veterinary Studies at the University of Edinburgh, Scottish Gamekeepers Association and the Scottish Wildlife Trust. Their mission statement expresses their views that captive breeding of wild-caught animals may be necessary to preserve the Scottish wildcat's unique traits, while acknowledging that those animals may not all be genetically 'pure' wildcats.

Wildcat Haven is a privately funded enterprise that focuses on *in situ* conservation rather than captive breeding. Its priority is establishing areas of wildcat habitat ('havens') that are entirely free of feral and hybrid cats so as to improve the survival chances of the wildcats that live in those regions. It too uses genetic and pelage scoring and TNR schemes to further its goals. Its first haven covers some 5366 square kilometres in Ardnamurchan and the surrounding areas, and in eastern Scotland a second is now being established. The organisation's website lists a comprehensive series of other goals too, including working with landowners, developing research and education programmes, and promoting ecologically friendly wildcat tourism.

Exploring the detailed websites of both organisations reveals that they are, unfortunately, at odds with one another over some factors of wildcat conservation, despite a shared ultimate goal. No field of human endeavour is conflict-free, but this serves to illustrate how complex the wildcat story is. The most thorny point is to do with genetic integrity and how much it matters.

The Q-test has shown us that no Scottish wildcats currently in captivity are 100 per cent wildcat, though some come very close. This raises the question of how things are in the wild – whether there are really any pure wildcats left at all. Through their genetic tests on wild-living wildcats, Scottish Wildcat Action concludes that free-ranging cats in Scotland are a 'hybrid swarm' with complete genetic mixing of Scottish wildcat and domestic cat, forming a smooth continuum of hybrids. No tested animal, in the wild or in captivity, has ever been found to be 100 per cent wildcat. If any do remain, their numbers will be tiny – as we've seen, estimates range from 35 to up to 2,000, with between 115 and 314 the estimate used by Scottish Wildcat Action. But it doesn't really matter what the true number is when we are talking about numbers of no more than 2,000, and when there are so many thousands of hybrids and feral cats out there. It would simply be impossible to find and catch enough entirely pure free-living Scottish wildcats to establish a viable and sufficiently genetically diverse captive-breeding population.

Compromising on genetic integrity is a bitter pill to swallow but a necessary one. As Christopher Clegg notes in his book *The Scottish Wildcat*, in essence, what we need to do is the reverse of what has been happening to our Scottish wildcats ever since domestic cats came onto the scene: water down the domestic cat DNA within wildcat populations (captive and free-ranging) to nothing or as near to nothing as we can manage.

The situation for Scottish wildcats in the wild is not wonderful right now, but it is better than it has been in decades. Persecution is minimal, and the numbers of viable feral cats and hybrids in the most likely wildcat areas are being reduced through TNR. The captive population continues to grow and its genetic integrity continues to improve, building a fund of animals that could be the beginning of a reintroduction. There's certainly plenty to feel optimistic about. Perhaps the Highland tiger can be saved after all.

Speyside, 2018

When I finally see the Scottish wildcat, after all these years of waiting, I suppose it's almost inevitable in a way that it should feel like an anticlimax. It's just suddenly there, when I least expect it, and I almost don't notice it at first. I don't even have binoculars or a camera at hand – they're both in my bag, so I grab my mobile phone, fumble to activate the camera, and manage one quick shot before the cat runs out of sight. I zoom in and the badly pixelated image shows just what I want to see – a clearly ringed, black-tipped tail with no dorsal line, and tawny flanks with what appear (as far as I can tell) to be unbroken fine stripes. It doesn't have any white markings. Oh, wait, it does. One white forepaw. So, despite that great-looking tail, this is not a real wildcat.

I already know this, I must admit, because I haven't met this cat in a forest or on a moor. In fact it was running across the forecourt of a petrol station on a side road in Grantown-on-Spey – only five minutes (as the cat trots) from the extensive forests south of town, but no wildcat would seek out the feel of concrete under its paws. Nevertheless, it is still the absolute best-looking wildcat-like cat I've ever seen outside of the zoos, and I stand there on the pavement feeling a growing sense of bewilderment.

It's March and I'm here in Grantown for a wildlife book festival at the lovely Grant Arms Hotel. I've arrived early and booked myself a room nearby for a few nights – a chance (my last) to spend time in wildcat country as I draw the threads of this book to a close. The house I'm

staying in backs onto Anagach Woods, the expanse of
pine forest that lies between the town and the Spey. The
woods are quite busy with visitors in the parts closest to
town, but they become wilder and more silent the
further you go. I have spent some time working in the
front room of the house but the garden's full of bird-
feeders, and the constant comings-and-goings of siskins,
coal tits and red squirrels is very distracting. Now I'm
heading for the bus stop to go down to Boat of Garten.
As I wait at the stop, I look again at my nearly-a-wildcat
photo. I wonder what its origins are. Is it feral or is it
owned? Has it been neutered? How far back up its
ancestral tree would we have to climb to find any true,
pure Scottish wildcats in its history – one generation?
A dozen?

I have lunch in the cafe here with Louise, who until
recently was working as staff naturalist at Aigas Field
Centre, not far from here. Aigas is a tour company but
it's also an important part of the Scottish wildcat
conservation effort here in Speyside. To quote their
website: 'Since 2011, Aigas Field Centre has been
contributing to an important national programme to
bring back the Scottish wildcat from the edge of
extinction. We are working in partnership with Scottish
Wildcat Action, which includes efforts to protect what's
left in the wild and a conservation breeding programme
led by the Royal Zoological Society of Scotland (RZSS).'
Aigas is one of five captive-breeding centres in the
Scottish Wildcat Conservation Action Plan. Louise's
work included care of several breeding pairs of cats living
in enclosures at Aigas, and now she is involved in camera-
trapping surveys.

I show Louise my awful photo of the forecourt cat
and she nods. 'Lots of the pet and feral cats around here
are part wildcat,' she tells me. Her charges at Aigas,
though, included some of the purest and wildest of
wildcats currently in captivity, and her work included
the nerve-wracking task of introducing tom to queen
in the hope of a successful mating but in anticipation
of possible violent fighting. I learn that Hamish, the
tom I met at the Highland Wildlife Park in 2013, has
sadly died recently but spent his last years at Aigas. I
remember his furious stare, but it's nice to hear that
actually he was of a reasonably mellow and playful
disposition for a wildcat, unlike another of Louise's
toms who bit her hand very badly through the thickest
of gauntlets when she had to catch him for an enclosure
move. Yet she says (as other wildcat keepers have told
me, too) that she didn't have a problem entering any of
the cats' enclosures under normal circumstances. Those
old tales of wildcats flying at human faces, sometimes
unprovoked and with claws out ready to rip, tear and
blind, would seem to be exaggerations. Either that or
the Scottish wildcat of today is a milder creature than
its ancestors.

Louise tells me of the ins and outs of wildcat care.
Unlike zoos, the Aigas facilities are closed to the public
in order to keep things as natural and undisturbed as
possible for the cats. The enclosures are furnished with
the kinds of things you might find naturally in wildcat
habitat: logs and boulders and living grass and bushes.
The birthing dens are designed to resemble natural
shelters as much as possible. The cats' diet includes whole
dead rabbits, quail and any salvageable roadkill the staff

happen to come across – and they supplement this by catching the occasional wild rodent that finds its way into their enclosures. To replicate wild conditions, once a week there's no food given at all (free-living wildcats would not necessarily manage to kill and feed every day). Then it's a matter of watching the females like a hawk for when they come on heat in winter, and making sure that the chosen males are introduced to them as soon as they show signs of receptivity.

Louise is encouragingly positive about the future of the wildcat despite some of the deeply unhelpful attitudes she's encountered while trying to get local people engaged in the drive to neuter and vaccinate domestic pets. When I leave her, I feel optimistic too, and take a long stroll through Abernethy Forest, seeing no wildcats nor anything much else but still enjoying the forest for exactly what it is.

The next day, I head into Anagach Woods and push northwards to get away from the busy trails close to Grantown. I'm about 5km in, and it's been ages since I saw a fellow human. I've also, excitingly, passed two signposts – the first saying 'Main Caper Area' and the second, a little further on, 'Core Caper Area'. According to my hosts, these woods are still home to a good population of capercaillies, and in my mind wildcats and capers go together – neither of them would put up with living quarters that suffered too much human disturbance. I've slowed down almost to tiptoeing pace and every time there's a good space between the trees I pause to lift up my binoculars and scan the branches (for capers) and the ridged and rolling ground (for cats). At one of these stops I glimpse a big shape skimming over the treetops and reflexively swing my camera up at it. To my surprise I find I'm photographing a red kite, but I learn later on that these lovely raptors are becoming more and more numerous in this area – wanderers from the Black Isle reintroduction scheme.

The path I have chosen takes me through the 'core caper area' and out the other side without any sign of a caper to show for it. The woods are beautiful though, and when I emerge from them it's into more beauty: the path runs through hedge-lined meadows with the Spey chuntering away in the middle distance. Out here I can hear wading birds. There's the excitable piping of oystercatchers and the beautiful wild warblings of curlews, and I think I can hear a redshank too. A lapwing lifts up from the rough grass and begins its ecstatic rolling display flight, rising and then tumbling upside-down, whooping like a toy trumpet all the while as it flashes white and black. Then its iridescent wings catch the light and it becomes momentarily multicoloured. I reflect that, not so long ago, you could walk in any area of low-lying farmland and see this kind of thing, but intensive agricultural practices have made it more and more difficult for waders to carve out a living. Nature reserves buck the trend, managing farmland in wader-friendly ways, but the land area they cover is tiny and simply not enough. It's so good to see these birds out and thriving in 'normal' countryside.

I walk back along the riverside, checking out the sky frequently in the hope of spotting an early migrant osprey on its return flight from Africa. I don't see one, but I do see dippers and at the bridge there are grey wagtails too – dazzling, dapper birds with Day-Glo yellow undertails that they flash frequently and gleefully as they flit from rock to rock, searching for waterflies. A little way along the path, the forest spills down almost to the river's shore, and here on a bare bank I find a mass emergence of mining bees. They are clambering out of their holes and blundering awkwardly about on the sandy earth – each grain of sand is a tennis ball-sized obstacle to their little feet. I think there are two species at first: a slim, stripy one and a fatter, furrier one. But then I remember the bits and pieces I know about bees and realise that I'm probably looking at males and females, respectively, of the same species. I am not sure of the species, though, so I take some photos to check out later.

The path eventually steers me back away from the river and up through the woods. This is the busy end and there are plenty of people out walking their dogs now. Overhead, siskins wheeze and I just pick up the soft twitter of a crested tit but I can't find it. I return to the house but head out again that evening, this time aiming for the small lochan that my map shows on the southern edge of the forest.

I'm picking my way along a thin pathway when I see the white bum of a grazing female roe deer ahead of me. She lifts her head to look at me and twitches her great petal-shaped ears in my direction, but she doesn't move away until I keep walking and get very close to her, whereupon she ambles up the slope next to the path and joins another couple of does. They all watch me go by, and show no particular sign of alarm. These deer are much less timid than those I've met in Abernethy Forest. I recall seeing photos of other wild mammals that have lost most of their fear of humans – pine martens skip-dancing across people's porches in broad daylight, foxes being handed chips by cabbies at my local station car park. I contemplate whether this habituation is something we'll see more and more, and I wonder why wildcats seem to be immune to it.

The path opens up and there's the lochan. Now I know why I've seen a few black-headed gulls flying over the house – they live here, just down the road. The lochan is a wide circle of shallow water, well vegetated at the fringes, with some weedy islands that look to me as though they could provide a safe mooring for a gull's nest. There are four pairs (well, I can't be sure they are pairs, but there are eight birds) of black-headed gulls here, a couple on the water and the others wheeling lazily about in their customary manner, gliding low then rising to describe a big circle around the margin of the lochan. They look pristine in their fresh breeding plumage. Whenever I see gulls, I feel a primal longing for seascapes, even though black-headed gulls will happily live and breed miles inland. On the lake are also mallards, which have hastened to the shore in front of me in a manner that suggests they are used to visitors bearing bread crusts. I wish I could return in summer and see what kinds of dragonflies would be exploring the margins of this beautiful little lochan.

I take another long walk the next day, heading north again and into the hills. That evening it's time to move over to the hotel and get ready for the book festival, where I'll be delivering my talk the next morning. Suddenly I'm among people, among friends, but I spend time alone, practising the talk I'm going to give and trying to keep my nerves under control. I'm grateful that

I'm the second of the many speakers – it'll be done and dusted soon and I can relax.

Public speaking scares me witless, but once I'm there and looking at the small but smiley audience, with my PowerPoint presentation safely loaded up and ready to go, it's like being at the top of the highest point on a rollercoaster. It's still scary as hell, but I'm riding the rails of my preparation and it's a fast and unstoppable ride to the finish. They laugh when they're supposed to, they ask wonderful questions, their warm applause cheers my heart, and it's over inside an hour. I sign some books, do some informal chat, but then I'm left alone and I head back up to my room. I put on my running gear and run like a woman possessed, out of the hotel, down the first side road and straight into the woods.

The trails, padded with fallen pine needles, feel deliciously soft and springy underfoot. The intense flood of relief powers me onwards. But one long hilly climb drains away my nervous energy completely and I pull up, catching my breath. I've come out into an area I've not explored until now, a big clearing with heather underfoot and astonishing views across the valley. There's the forest downslope, falling away to the bright ribbon of the river, and the meadows all around. Beyond are rising hills, marbled by shifting clouds' shadows, and further still loom the hulking misty outlines of the mountains. Time to stop and to drink it in. I don't get to walk in beauty like this every day.

Although my impromptu run means I miss the talk right after my own, I attend the others that day. Some of the speakers attending this event are household names (if your household is one of wildlife enthusiasts), some are friends, some are both. The talks are enjoyably varied – from avian anatomy to wild food foraging, via ravens and springtime and the unsung islands that circle our coastline. I am rapt and inspired and greatly encouraged. Working and writing alone every day is a fight against isolation and the pessimism that can bring. I've been trapped in my home and in my head too much lately, and being among other nature-lovers and hearing their impassioned words is exactly what I need.

When the afternoon's talks are over, I go into the woods again but this time I'm not alone – a little gang of us authors are walking together to find the bird- and squirrel-feeding station that's set up here by the hotel. It turns out I have walked right past it twice without noticing it. We stop on the path and watch a red squirrel chasing another down the tree trunk to where their box-shaped feeder hangs. Nearby is a tube-shaped bird-feeder that is thronged with coal tits. We're holding out for crested tits, though, and after ten minutes' wait one shows up. There's something extra satisfying about being out watching wildlife with these people – putting our words into actions.

The evening talk that day is by Mark Cocker, who discusses his book *Our Place: Can We Save Britain's Wildlife Before It Is Too Late?* He pulls no punches – much of what he has to say is sobering, shocking, frightening. But all of us in the room know it – these are the truths we have to face. Our place, our countryside, has been denatured, wounded and disfigured by all we have done to it over the centuries and we've already lost so much. It often seems as though the only way to see wildlife now is on nature reserves. They are marvellous but they're not

enough – they're tiny oases within the vast and sterile
landscape of the wider countryside. It would be easy to
throw up our hands and say the situation is hopeless, but
Mark's not for this and neither are we. He starts to talk
about the waders he's seen here in the meadows, down
by the Spey, the tumbling lapwings. I remember them
too and my mood rises, as does that of the whole
room. This landscape, right here, is proof that 'ordinary'
countryside can still teem with wildlife. We know the
way – we just need to harness the will.

I go to bed that night, my last night here, with a head
full of hope. Saving the Scottish wildcat is a microcosm
of all the work that faces us to restore Britain as a fit
place for wildlife, as well as for people. But this is one of
those times when I truly feel that we can do it.

I am reluctant, as ever, to leave the Highlands, but I've
an ex-feral cat of my own to get back to. While I've been
away, a friend has been feeding Sookie for me, but Sookie
doesn't take to strangers and has spent the whole week
hiding under the bed. It takes me ten minutes to coax
her out, but after that our reunion is joyous. As I massage
her stripy head and think about the hybrid cat I saw in
Grantown-on-Spey, I reflect on how, when it comes
down to it, it's a love of cats – all kinds of cats – that is
our strongest motivation in the fight to save the Scottish
wildcat and, at a stroke, help make life better for pet and
feral cats as well.

CHAPTER EIGHT

An Uncertain Future

Imagine a forest walk, a few decades hence. Your guide leads you between tall, twisty pine trees along a waymarked trail. The trees open up to grazing moorland and you follow your guide to a small, unobtrusive hide. You get settled in, prop your elbows on the shelf and rest your binoculars in your hands, and wait as the light slowly fades across your view of forest edge, rough tufty grassland and boulder-strewn slope. Excited, looking from rocks to rushes and back again, you imagine movement everywhere. A hunched shape that shifts minutely. A glimpse of a striped flank. The flick of a banded tail. Then you see real and definite movement – a tawny, striped animal appears in the mouth of a dark

space between the stacked boulders way ahead. You hold
your breath as you watch it stretch, fore and aft, and
amble down towards you. Then, on its heels, two
miniature versions of itself come skipping behind. The
wildcat family heads off towards the forest edge and you
watch them go, feeling astonished, overwhelmed and
grateful.

That's the dream, and if the ongoing conservation
plans achieve their goals, that's the reality in store for our
future selves and future wildcats. The alternative – the
worst-case scenario – is that we reach a point where we
have to confirm that nothing that can be called a Scottish
wildcat survives any more, only ferals and obvious
hybrids, with each new generation of hybrids carrying a
little less Scottish wildcat than the last. The Scottish
wildcat as a living, complete entity would have ceased to
be; it would exist only as a dispersed scattering of genes
hiding in the cells of a dwindling number of slightly
wildcat-like feral cats.

Right now, we're a long way from either of these
possibilities, but without current conservation efforts
we'd already be well on our way to that worst-case
scenario. And it wouldn't take too much of a crisis to put
us back on that path again, so we cannot afford to be
complacent. Conservation work is costly and often the
first thing to go when budgeting belts are tightened. You
only need to look to other countries with less in the
coffers than we have here in order to see that the will to
save a species isn't always enough – if the funding can't
be found, the work can't be done and the species can't
be saved.

Let's revisit our list of all 40 wildcat species in the world – this time with their IUCN conservation status. The threat levels that the IUCN uses for species that are extant in the wild are, in escalating order of seriousness, as follows: Least Concern, Near Threatened, Vulnerable, Endangered, Critically Endangered. Only Least Concern is 'safe'; all the other categories reflect a real risk of extinction in the not-too-distant future, the level of risk being assessed according to a range of factors. These include their estimated population size at present, the extent of habitat available to them, the type and seriousness of the threats they face, and whether they are currently declining (and if so, how rapidly).

Lion *Panthera leo*: Vulnerable
Jaguar *Panthera onca*: Near Threatened
Leopard *Panthera pardus*: Vulnerable
Tiger *Panthera tigris*: Endangered
Snow leopard *Panthera uncia*: Vulnerable
Sunda clouded leopard *Neofelis diardi*: Vulnerable
Clouded leopard *Neofelis nebulosi*: Vulnerable
African golden cat *Caracal aurata*: Vulnerable
Caracal *Caracal caracal*: Least Concern
Serval *Leptailurus serval*: Least Concern
Pampas cat *Leopardus colocola*: Near Threatened
Geoffroy's cat *Leopardus geoffroyi*: Least Concern
Güiña *Leopardus guigna*: Vulnerable
Southern tiger cat *Leopardus guttulus*: Vulnerable
Andean mountain cat *Leopardus jacobita*: Endangered
Ocelot *Leopardus pardalis*: Least Concern
Northern tiger cat *Leopardus tigrinus*: Vulnerable
Margay *Leopardus wiedii*: Near Threatened

Borneo bay cat *Catopuma badia*: Endangered
Asiatic golden cat *Catopuma temminckii*: Near Threatened
Marbled cat *Pardofelis marmorata*: Near Threatened
Canada lynx *Lynx canadensis*: Least Concern
Eurasian lynx *Lynx lynx*: Least Concern
Iberian lynx *Lynx pardinus*: Endangered
Bobcat *Lynx rufus*: Least Concern
Puma *Puma concolor*: Least Concern
Cheetah *Acinonyx jubatus*: Vulnerable
Jaguarundi *Herpailurus yagouaroundi*: Least Concern
Leopard cat *Prionailurus bengalensis*: Least Concern
Sunda leopard cat *Prionailurus javanensis*: Least Concern
Flat-headed cat *Prionailurus planiceps*: Endangered
Rusty-spotted cat *Prionailurus rubiginosus*: Near Threat-
    ened
Fishing cat *Prionailurus viverrinus*: Vulnerable
Pallas's cat *Otocolobus manul*: Near Threatened
Chinese mountain cat *Felis bieti*: Vulnerable
Jungle cat *Felis chaus*: Least Concern
African wildcat *Felis lybica*, including the domestic cat
    F. l. catus: Least Concern
Sand cat *Felis margarita*: Least Concern
Black-footed cat *Felis nigripes*: Vulnerable
European wildcat *Felis silvestris*, including the Scottish
    wildcat *F. s. grampia*: Least Concern
(IUCN 2018. The IUCN Red List of Threatened
    Species. Version 2018-2. www.iucnredlist.org.)

So that makes 15 species that are of Least Concern – the
biggest single grouping. However, with seven Near
Threatened species, 13 Vulnerable and five Endangered,
there are 25 altogether that face some degree of threat.

That's nearly two-thirds of all the world's cat species looking down the barrel of extinction, including all of the big cats. There's another statistic to consider alongside this, too: the population trend. And if you thought the IUCN categories made for depressing reading, you may want to look away now. Of all these species, only one is increasing: the Iberian lynx, which has been pulled out of its previous Critically Endangered category thanks to concerted conservation efforts, including captive-breeding projects. Of the rest, six have a stable population (Geoffroy's cat, serval, Canada lynx, Eurasian lynx, bobcat and leopard cat), and for two more (caracal and sand cat) the population trend is not known. The other 31 are all declining, which means they're on their way to a higher-threat category if things don't change for them.

Those five Endangered species are in trouble for a variety of reasons. The tiger – flagship animal of conservation as a whole – is most threatened by deliberate hunting to satisfy an enthusiastic black market for its body parts with their rumoured medicinal benefits. It's hard to get your head around this but, as the wild tiger population has dwindled, this demand has only increased and the financial rewards for poachers are tremendous. Tigers are also fighting it out for space with growing human populations, and where people and tigers live in close proximity, tigers will kill livestock (and occasionally people), for which they are then killed in retaliation.

The rare and little-known Andean mountain cat is found in the high mountainous regions of the central Andes – the sort of terrain that should, in theory, keep it somewhat safe from human disturbance, but sadly this cat is targeted disproportionately by local hunters. It is considered too much of a threat to poultry and other small livestock, and it is also used for medicine and in ceremonies and festivals. The fact that it is legally protected in all of the countries where it occurs has, to date, not done much to diminish the problem. Heavy hunting of its preferred prey species is another of the factors that have brought its estimated wild population down to fewer than 1,400 individuals.

The Borneo bay cat lives only on the island of Borneo. A red- or grey-furred, long-bodied and puma-like animal, though only the size of a domestic cat, it occurs in dense forest, and dense forest is a fast-dwindling habitat on its home island. It is vanishingly rare, almost never seen in the wild, and (so far) does not thrive in captivity, so captive breeding may not be a valid way to increase its numbers (no more than 2,200, by the IUCN's assessment).

I remember reading of the flat-headed cat in one of my favourite childhood books, a field guide to the world's felines (I was a very geeky child). Its picture fascinated me – its small and low-set ears, coupled with bulging dark-amber eyes, gave it an almost lemur-like appearance. This weird little cat occurs on Borneo, Sumatra and the Malaysian Peninsula, and lives in swampy forests. It is very rare, its habitat is dreadfully pressured, and there are probably fewer than 2,500 left now. It has legal protection, but actual proactive conservation efforts are all but non-existent.

The only Endangered species – in fact, the only wild cat species – that is actually doing well at this moment in time is the Iberian lynx, and the success of its conservation scheme has both inspired and informed those working on Scottish wildcat conservation. This is a beautiful lynx, smaller and sleeker and more gracile than the Eurasian lynx, with particularly extravagant cheek frills and ear tufts. It formerly ranged across Spain and Portugal but today occurs in just two widely separated locations in southern Spain. I went to one of them, the Sierra de Andújar national park, in January 2013 in the hope of seeing it – and also perhaps seeing European wildcats, which also occur there.

I knew from my explorations of dense forests in northern Europe how small the chances are of bumping into a Eurasian lynx. It seemed intuitively right that to see the far rarer Iberian lynx would be even more difficult, and as I stood on a roadside, surveying a huge sweeping valley of rocky scrubland, my natural optimism did founder a little. But then I turned away from the vista to glance back down the way we had come, and my eyes met those of a real, live, perfect Iberian lynx, casually crossing the road a few metres away. It paused to scrutinise me coolly, then turned back the way it was going, reached the edge of the road and headed away downhill, quickly vanishing among the bushes and boulders.

Somehow I'd had the presence of mind to take a couple of pictures, even though the lynx was far too close to me to fit in the camera frame. They are among the closest-range photos I've ever taken of any wild mammal. I couldn't quite believe my luck, but within the hour I had seen and photographed another two lynxes, from the same spot. Those three cats represented nearly 2 per cent of the world's entire population of Iberian lynx at the time, estimated at just 156 individuals in 2012 by the IUCN (although that number is a threefold increase from the population in 2002).

This cat's decline has a key cause that's different again to the other four Endangered species: a catastrophic decline in its key prey species. Iberian lynxes eat rabbits – almost to the exclusion of anything else. Rabbits are native to Iberia and north Africa, and while their various introduced populations around the world are mostly thriving, here on their actual home turf they have suffered huge losses due to overhunting, habitat loss, and most importantly the impact of myxomatosis and rabbit haemorrhagic disease. Loss of rabbits meant loss of lynxes (inevitably lynxes were also persecuted, but it was the decline in the rabbit population that really did for them). Lynx conservation schemes, therefore, have had to also be rabbit conservation schemes, focusing on increasing rabbit numbers through reintroductions and habitat improvements. By imposing traffic calming measures, the project has also helped reduce the number of lynxes killed on the roads, and conservationists have also used captive breeding, reintroduction and relocation (between the two core sites). All of this work has helped the Iberian lynx recover its numbers enough to be pulled out of the Critically Endangered category into merely Endangered – there are more than 400 living wild today. But you don't have to be any kind of biologist to know that a few hundred individuals is still a perilously tiny population. It will take many more years of concerted effort before the Iberian lynx's future can be regarded as anything like secure.

So the Scottish wildcat is far from alone in being a cat on the edge, but its situation is probably even worse than any of the above species'. The IUCN doesn't routinely assess subspecies, so it has no individual entry for the

Scottish wildcat. However, at the time of writing, its account for *Felis silvestris* does mention wildcats in Scotland as follows:

> *Recent estimates have varied between 1,000 and 4,000 (compared to 1.2 million feral cats in Britain), but as few as 400 cats which meet morphological and genetic criteria for being the furthest from the domestic group may survive (Macdonald et al. 2004, Battersby 2005, Kitchener et al. 2005, Macdonald et al. 2010). If so, this population would be Critically Endangered (Kitchener et al. 2005).*

The dangers facing the Scottish wildcat are well understood and, although they are difficult problems to tackle, there is a plan going forward that *should* work. The main worry and debate is really whether it is already too late, with the *grampia* genome perhaps already too diluted by domestic cat genes. This is almost more of a philosophical debate than a practical one. That even the 'best' wildcats carry some domestic genes seems beyond dispute now, so it's our decision to make as to how much wildcat DNA is needed to deem a cat a real or even just a 'good enough' wildcat. Scottish Wildcat Action sets a high bar when selecting cats for their captive-breeding project. In their paper 'Wildcat Hybrid Scoring For Conservation Breeding under the Scottish Wildcat Conservation Action Plan', the organisation explains their decision-making on this as follows:

> *Any programme to bring 'wildcats' into a conservation breeding programme will have to set a threshold (based on judgement rather than a clear biological distinction) that*

balances the wish to preserve the genetic diversity encapsulated in the apparently non-pure wildcats from Scotland as part of a Scottish Wildcat Conservation Breeding Programme, versus the desire not to be too inclusive of domestic cat genes (and associated traits). Set the purity bar too high and the risk is that good wildcat genes are excluded, 'good-ish' cats are excluded as hybrids, and the population accepted into captivity is so small that it will experience a high level of inbreeding. Set the bar too low and we end up breeding something that is only slightly better than the situation in the wild.

The captive-breeding project is well established and running smoothly, and the stock of high-scoring cats grows each year – at the time of writing, there are about 80 of them. They are kept in such a way that natural behaviour is supported and encouraged as far as feasible. So how far away are we from a situation where it will be possible to start introducing some of these animals into the wild? A little way off yet, it turns out. There is more to do, and a lot of it starts with ordinary people who happen to live in wildcat country. Many more feral cats need to be neutered in the core wildcat areas, and more people need to be reached through educational programmes that encourage neutering and vaccination of pet cats.

The public's involvement in documenting observations of free-living cats of all kinds is also invaluable. Scottish Wildcat Action's new #GenerationWildcat campaign, initiated in 2017, is key to this. It calls on the public – including outdoor enthusiasts, farmers and gamekeepers – to join the fight to bring the Scottish wildcat back from the edge of extinction. Dr Roo Campbell, SWA Project Manager, says: 'The time to save the Scottish wildcat is now. We are almost certainly the last generation who has a realistic chance of saving this iconic species from extinction in Scotland. Wildcats here face three key threats: hybridisation with feral domestic cats, disease and accidental killing.' The project encourages everyone living in the areas where wildcats occur to:

1 report sightings of wildcats, living or dead (these can be logged through Scottish Wildcat Action's website or via the downloadable 'Mammal Tracker' app – dead wildcats should be photographed carefully from all angles so that a full pelage score can be made)
2 report sightings of feral cats
3 make sure their own cats are neutered – this is particularly crucial in the priority areas of Morvern, Strathpeffer, Strathbogie, northern Strathspey and the Angus Glens.

It also asks that farmers keep their farm cats healthy and neutered, and that gamekeepers use cage traps instead of lamping/snaring to control feral cats, so that any wildcats caught in error can be saved. Perhaps the most user-friendly element of their advice, though, is a comprehensive guide on how to set up one's own trail cameras to look for wildcats across Scotland.

Surveys of habitat quality, prey abundance and threat level must also be carried out to identify the places where wildcats are most likely to survive and thrive. The captive wildcats are precious – none of them should be sent out into the wild until we are reasonably certain that they have a good chance of surviving and breeding. When release does happen, there will need to be ways to monitor the cats as they make their way in the wild. Satellite tagging works beautifully to track individual wild animals in some circumstances, but because the batteries that power these devices are solar-charged and wildcats are mainly nocturnal, other ways may be better for them.

Another part of the process is habitat creation and enhancement. Scottish Wildcat Action has placed tracking collars on a few free-living wildcats over the last few years to discover how they use their habitats, and this has helped identify ways of making wild places more hospitable to wildcats. Sometimes this means clearing patches of forest, in order to encourage the mosaic of tree cover and open ground that wildcats prefer. It also means creating artificial denning sites as the lack of a good den can render otherwise perfect habitat no good for wildcats.

It's exciting to read about the future plans for Scottish wildcat conservation – once you have made the mental leap that they all involve wildcats that are not quite pure. But, to glance backwards for a moment, the pure Scottish wildcat was probably doomed from the moment the first domestic cats set paw on our soil, long before our own wildcats were actually (just) Scottish. This begs the following question: what about wildcats elsewhere in Britain? Might we some day seek to return wildcats to England and Wales and even Ireland? And, if so, should they be Scottish wildcats or European wildcats? Derek Gow – an expert on mammal reintroductions and a key figure in the projects to return European beavers to our waterways — is in favour of introducing European wildcats to parts of Britain, highlighting that they could help control grey squirrel numbers and should have no negative impact on any native species. However, they would be vulnerable — just as Scottish wildcats are today — to hybridisation with domestic cats, and this risk would have to be mitigated. The question of what would happen if reintroduced European wildcats (subspecies *silvestris*) met native Scottish wildcats (subspecies *grampia*) is also a vexed one.

Reintroduction of the Eurasian lynx could happen within the next few years. This, too, could impact Scottish wildcats through competition rather than hybridisation — research on mainland Europe shows that European wildcats don't fare so well in areas that hold lynxes. However, contact between the two is an unlikely scenario in Britain (at least, not for many generations) as lynxes will not be released in key wildcat areas.

The general principle of 'rewilding' has gathered great momentum as an ideological movement over the last couple of decades. The charity Rewilding Britain sets out the movement's principles and goals on its website as follows:

> *Rewilding is the large-scale restoration of ecosystems where nature can take care of itself. It seeks to reinstate natural processes and, where appropriate, missing species — allowing them to shape the landscape and the habitats within. Rewilding encourages a balance between people and the rest of nature where each can thrive. It provides opportunities for communities to diversify and create nature-based economies;*

for living systems to provide the ecological functions on which we all depend; and for people to re-connect with wild nature.

We believe in four principles of rewilding.

1 *People, communities and livelihoods are key. Rewilding is a choice of land management. It relies on people deciding to explore an alternative future for the land and people.*
2 *Natural processes drive outcomes. Rewilding is not geared to reach any human-defined optimal point or end state. It goes where nature takes it.*
3 *Work at nature's scale. Rewilding needs sufficient scale so that nature can reinstate natural processes and create ecologically coherent units.*
4 *Benefits are for the long term. Rewilding is an opportunity to leave a positive legacy for future generations. It should be secured for the long term.*

Rewilding projects cover reintroductions, habitat enhancements, restorations of ancient landscapes, more wildlife-friendly management of farmland, and much more. By their very nature they are on a grand scale. Reading about rewilding is hugely inspiring. One project, on the Knepp estate not far from where I live, has involved transforming an unprofitable and wildlife-depleted farm to a spectacular mosaic of natural habitats, which have attracted rare wildlife such as turtle doves, nightingales and purple emperors – and it remains a farm, too, but now it is a profitable farm. The same can be done anywhere, given the will and the work. Wildcat country is ripe for rewilding and the process is already beginning. The return of the wildcat itself is not the only goal, though. It's just one part of a wider and wilder story.

The Scottish wildcat's story is a singular tale of an exceptional animal facing an unusual and intractable range of problems, but it is still a component of the wider and wilder world. Only if we place this cat in context can we really see its problems and their solutions.

When I began this book I didn't expect to be finishing it with a real sense of hope, but to my surprise I've learned that hope is indeed an option – the only option. The Scottish wildcat, with the help of all those who care for it, is fighting as fiercely as the wildcats of legend to survive and reclaim its lands. I, for one, wouldn't bet against it.

Acknowledgements

I owe many people a debt of gratitude for the help they have provided over the time I've spent working on this book.

Thank you to Jim Martin for commissioning this project and Julie Bailey for seeing it through the later stages – and to both of you for your immense patience with me through the hold-ups and tricky moments. I'm grateful to Charlotte Atyeo for copy-editing my text so well, and to Liz Drewitt for proofreading. The cover artwork is the work of Ian MacCulloch – I fell in love with his style after he created the artwork for my previous book, *The Way of the Hare*, and I am so pleased that he agreed to work with us again (and that he listened patiently to my rather quirky ideas about how I'd like it to look).

I talked to a number of wildcat experts in the process of researching this book, and read the words of many more. I am very thankful to each of them for all they have done and continue to do to further Scottish wildcat knowledge and conservation. In terms of the written word, the beautiful book *The Scottish Wildcat* by Christopher Clegg, and the extensive and detailed websites of Scottish Wildcat Action (www. scottishwildcataction.org) and Wildcat Haven (www. wildcathaven.com) have been particularly helpful resources. I'd urge everyone reading this to check them out.

Thanks to everyone who provided accommodation and invaluable local tips for me during my stays in Scotland – Polly and Ross Cameron at Dell Cottages in Nethy Bridge, Heatherlea at the Mountview Hotel also in Nethy Bridge, Kilchoan campsite in Ardnamurchan, and Donald and Libby at Hearthside and the Grant Arms Hotel in Grantown-on-Spey. Thanks to my fellow authors for making my last trip to Speyside so enjoyable, and also to my guides and fellow guests on the Heatherlea Outer Limits tour.

My friends have, as ever, offered support and tea and words of wisdom through the writing process and I am so grateful to you all. Some of you even came with me on a search or two or a visit to meet captive wildcats, while others made me jealous with your stories of seeing real wildcats for yourselves.

Thank you to Marion of RAIN and Kate of Streetkatz for helping me to get involved with feral cat rescue and to learn so much about the issues that surround it. And thanks to the cats themselves who have unfailingly entertained and enchanted me over the years. A special thanks to my most beloved formerly feral and still somewhat wild cat, who's learned to love me nearly as much as I love her and who has kept me company over this last year – Sookie, the original *gatita fiera*.

Finally, thanks to Alex. Free-faller and star, knower of words and befriender of *gatitas* – the vivid viverrid on my shoulder who helped me bring it all home. All the way to Mars and, yep, back here again. This one's for you.

Further Information

Books

Clegg, Christopher. 2017. *The Scottish Wildcat*. Shropshire: Merlin Unwin Books.
A fully illustrated and comprehensive study of the Scottish wildcat past and present, exploring its history and biology, and its conservation prospects.

Elder, Charlie. 2015. *Few And Far Between: On The Trail of Britain's Rarest Animals*. London: Bloomsbury Publishing.
Charlie Elder's quest to find the most threatened and elusive of British wildlife naturally included a search for Scottish wildcats, along with an array of other similarly enigmatic species.

Harris, Stephen and Derek Yalden. 2008. *Mammals of the British Isles: Handbook*. Fourth edition. London: The Mammal Society.
This book covers in great detail the ecology of all wild mammals present in the British Isles, with a wealth of useful data on historical and current populations and distributions, dietary breakdown, breeding biology and much more.

Tomkies, Mike. 2016. *Wildcat Haven*. Caithness: Whittles Publishing
Journalist turned conservationist, Mike Tomkies recounts the remarkable experience of rearing, caring for, breeding and eventually re-releasing a number of Scottish wildcats. He then went on to breed and release several litters of wildcats.

Documentary

The Tigers of Scotland. 2017. Directed by Leanne Gater. Produced by Wild Films Ltd. Available to watch on Netflix UK (as of going to print) or to buy from Amazon Prime and iTunes, 55 min.
This documentary, made with the help of conservation body Scottish Wildcat Action, takes us into the heart of wildcat country, and we meet the conservationists working on all fronts to save the Scottish wildcat, preserve its genetic integrity, and restore and manage its habitat.

Websites

Aigas Field Centre: aigas.co.uk
This study centre offers courses and learning holidays in the heart of the Highlands, and is part of the national breeding project for Scottish wildcats, in association with Scottish Wildcat Action and the Royal Zoological Society of Scotland.

Highland Wildlife Park: highlandwildlifepark.org.uk
This popular and strongly conservation-focused small zoo, situated in the Speyside area of Scotland, is a crucial part of the Scottish wildcat national captive breeding project.

International Union for Conservation of Nature: iucn.org

The IUCN is a global body focused on conservation of nature throughout the world. Its members include 219 states and government agencies, and well over 1,000 non-governmental organisations. It is involved with environmental research and protection, environmental law, education and individual species studies.

IUCN Red List: iucnredlist.org

This regularly updated list, published by the IUCN, presents data on the conservation status of nearly 100,000 species worldwide, with an assessment of threat level for each.

IUCN SSC Cat Specialist Group: catsg.org

This component of the IUCN's Species Survival Commission is focused on conservation of cats in the wild and maintains the IUCN Red List for all wild feline species. It has nearly 200 members across 62 different countries.

Lynx UK Trust: lynxuk.org

This trust, formed by a group of specialist scientists and conservationists, is working to secure a reintroduction of the Eurasian lynx, our only native wild cat besides the Scottish wildcat, in Scotland and England.

Panthera, Small Cats Program: panthera.org/initiative/small-cats

Panthera organises and promotes conservation plans for all species of wild cats – although its focus is on the big cats its work also includes developing conservation strategies to help small cat species.

Scottish Wildcat Action: scottishwildcataction.org

This partnership project is working towards monitoring and saving Scottish wildcats in the wild and building a sustainable captive population of pure or almost pure wildcats for future re-releases. More than 20 conservation organisations support its work. The website includes a wealth of resources, including full details of how genetic testing and pelage scoring is carried out on putative wildcats.

Small Wild Cat Conservation Foundation: smallcats.org

This organisation, founded by wild cat expert and conservationist Jim Sanderson, promotes the survival of the world's small wild cat species, and conservation of their habitats.

Wildcat Haven: wildcathaven.com

A privately funded project focused on identifying and protecting key areas of wildcat habitat, firstly by removing or neutering all feral cats and hybrids in these zones, and establishing buffer zones to ensure no further hybridisation will occur. The project is also involved in education and monitoring programmes.

Wildwood Trust: wildwoodtrust.org

This collection of native British wildlife, open to the public, includes Scottish wildcats which are bred here as part of the national captive breeding project. It is located close to Canterbury and Kent.

Index